FORTY
AUTUMNS

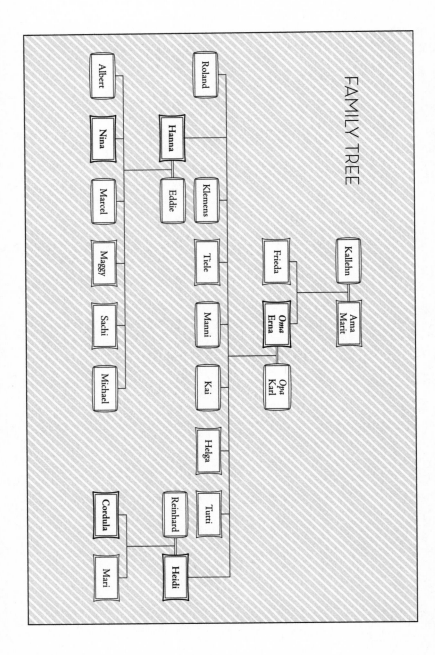

FAMILY TREE

FORTY AUTUMNS

A FAMILY'S STORY OF COURAGE AND

SURVIVAL ON BOTH SIDES OF

THE BERLIN WALL

NINA WILLNER

WM

WILLIAM MORROW

An Imprint of HarperCollins*Publishers*

for

OMA

HarperCollins books may be purchased for educational, business, or sales promotional use. For information please e-mail the Special Markets Department at SPsales@harpercollins.com.

A hardcover edition of this book was published by William Morrow in 2016.

FIRST WILLIAM MORROW PAPERBACK EDITION PUBLISHED 2017.

Designed by Bonni Leon-Berman

Library of Congress Cataloging-in-Publication Data
Names: Willner, Nina, 1961– author.
Title: Forty autumns : a family's story of courage and survival on both sides
  of the Berlin Wall / Nina Willner.
Description: First edition. | New York, NY : William Morrow, [2016] |
  Includes bibliographical references.
Identifiers: LCCN 2016038924 | ISBN 9780062410313 (hardback)
Subjects: LCSH: Willner, Nina, 1961—Family. | Germany (East)—Biography. |
  Women—Germany (East)—Biography. | Berlin (Germany)—Biography. |
  Berlin Wall, Berlin, Germany, 1961–1989—History. | Willner, Nina, 1961– |
  German Americans—Biography. | Women intelligence officers—United
  States—Biography. | Intelligence officers—United States—Biography. |
  BISAC: BIOGRAPHY & AUTOBIOGRAPHY / Personal Memoirs. |
  BIOGRAPHY & AUTOBIOGRAPHY / Military. | BIOGRAPHY &
  AUTOBIOGRAPHY / Historical.
Classification: LCC DD281.5 .W56 2016 | DDC 929.20943—dc23 LC record
  available at https://lccn.loc.gov/2016038924

ISBN 978-0-06-241032-0 (pbk.)

18 19 20 21   ov/LSC   10 9 8 7 6 5

Both now and for always, I intend to hold fast to my belief in the hidden strength of the human spirit.

—*Andrei Sakharov, Russian nuclear physicist and dissident*

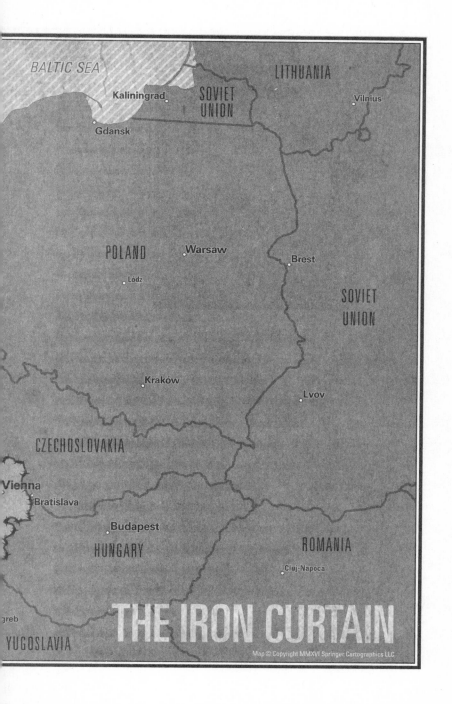

BALTIC SEA

LITHUANIA

Kaliningrad

SOVIET
UNION

Vilnius

Gdansk

POLAND

Warsaw

Brest

Łódź

SOVIET
UNION

Kraków

Lvov

CZECHOSLOVAKIA

Vienna

Bratislava

Budapest

ROMANIA

HUNGARY

Cluj-Napoca

greb

YUGOSLAVIA

THE IRON CURTAIN

Map © Copyright MMXVI Springer Cartographics LLC

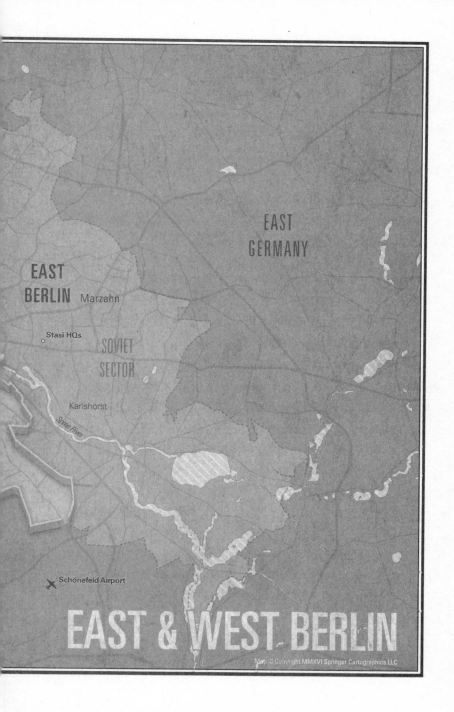

EAST
GERMANY

EAST
BERLIN Marzahn

Stasi HQs

SOVIET
SECTOR

Karlshorst

Spree River

Schönefeld Airport

EAST & WEST BERLIN

Map 3 Copyright MMXVI Springer Cartographics LLC

CONTENTS

FAMILY AND HISTORICAL CHRONOLOGY

YEAR	FAMILY	HISTORY
1945	Americans arrive in Schwaneberg Soviets take command of Schwaneberg	Red Army captures Berlin World War II ends. Germany divided into East and West. Berlin divided into East and West. Cold War begins
1946	Opa installed to teach Soviet doctrine Roland becomes teacher	Soviets occupy East, impose Soviet law Establish border demarcation and interzone pass control SED, Communist Party formed VoPo police and German border police established
1947	Kallehn helps Hanna to flee Hanna escapes and is forcibly returned	Marshall Plan helps rebuild West Zone and West Berlin Soviets strip East Zone Communist youth movement established
1948	Hanna's final escape	Currency reform Berlin Blockade/Berlin Airlift begins
1949	Hanna to Heidelberg Heidi born in East Zone Youth movement begins in Schwaneberg	NATO formed Berlin Blockade ends West Germany founded East Germany founded Prison system building Soviets test first nuclear weapon. Nuclear arms race begins. Group of Soviet Forces Germany (GSFG) established
1950	Opa takes stand for farmers against state	Ulbricht becomes leader of East Germany East German state confiscates private land Restricted areas established Stasi established. Begins to use fear, paranoia, intimidation, and terror tactics to control.

YEAR	FAMILY	HISTORY
1951–56		East Germany's Five Year Plan stresses high production quotas for heavy industry. Mass exodus of worker and intellectual labor force.
1952	U.S. Army hires Hanna	East-West German border sealed. Only Berlin remains open.
1953		Stalin dies. New Course.
		Construction of Socialism Plan
		Workers' Uprising. Riots suppressed by Red Army.
1954	Oma and five-year-old Heidi visit the West	Unauthorized departure from East Germany prosecuted by three-year prison term
1955	Family adjusting to police state control	Soviet Union declares East Germany sovereign
		Warsaw Pact formed
	Authorities harass Opa	Regime sees first results of silencing dissent and controlling population
		Normalization begins
		Cold War escalates
1956	Kai takes Jugendweihe oath to serve communism	NVA (East German Army) formed. With GSFG, ramps up to become battle-ready for conflict with NATO.
		Hungarian Revolution suppressed by Red Army
1956–63		Seven Year Plan marked by collectivization and nationalization of agriculture and industry
1957		Assisting in unauthorized departure from East Germany prosecuted by prison term
		Space race begins
1958	Hanna marries Eddie	
	Oma and Opa visit Heidelberg	
1960	Authorities continue to harass Opa	Cold War escalates
		Socialist Spring. Final handover of land for collectives.
	Kallehn dies	
		Continued hemorrhaging of East German labor force, mostly into open West Berlin

YEAR	FAMILY	HISTORY
1961	Nina born in United States	Berlin Wall erected
		Tensions increase between United States and Soviet Union
1962	Kai serves as border guard at the Berlin Wall	Peter Fechter is shot at the Wall
	Oma builds Family Wall	
1963		U.S. president Kennedy "Ich bin ein Berliner" speech
1963–70		New Economic System
1964	Heidi takes Jugendweihe	Berlin Wall fortified
1965	Opa denounced, forced to retire, expelled from Communist Party. Family relocated to Klein Apenburg.	
1966	Nina, age five, learns East family is trapped in a country they cannot leave	Infrastructure improvements: roads and apartment blocks
1968	Heidi marries Reinhard	Soviets crush revolt in Czechoslovakia
		East German regime launches sports program
1969		Détente begins
1970	Cordula born	East Germany enters Olympics
1971		Honecker becomes leader of East Germany
1972		West and East Germany establish diplomatic relations
1973	Hanna calls Oma in East Germany	Consumer socialism
		East German sports ramps up
1975	Family receives more letters and packages from Hanna	Wall upgrade; border reinforced
		Escape attempts, now almost suicidal, continue
1976		SS-20 missiles deployed in the Soviet Union
		Peak of growth in East Germany

YEAR	FAMILY	HISTORY
1977	Opa sent to asylum, "reeducated"	
	Kai dies	
1978	Oma dies	
	Albert visits East German family	
1979		Soviet Union invades Afghanistan. Détente ends.
1980	Cordula recruited into East German sports	Cold War tensions increase. Solidarity forms in Poland.
	Heidi and Reinhard build Paradise Bungalow	Economic decline; conditions deteriorate. More East Germans tuning in to West airwaves.
		Allotment garden plots given out
1982		Reagan's "war on communism"
		Intelligence collection at all-time high in East and West Berlin
1983–86	Nina works intelligence operations in East Berlin	U.S. Pershing missiles deployed in West Germany
		Reforger 83 and Able Archer 83 simulate conventional, chemical, and nuclear war in Europe. Soviets perceive possibility of NATO first strike.
		Tensions escalate
1984	Cordula takes Jugendweihe oath	
	Opa dies	
1985	Cordula enters national team and trains in East Berlin	Soviet leader Gorbachev begins reforms. Honecker opposed to change.
	Nina operational in East Berlin	U.S. Army Major Nicholson shot and killed while on USMLM mission in East Germany
1986	Hanna and Eddie visit Berlin	
1987	Cordula travels outside East Germany to compete in world championships. Prepares for Olympics.	Reagan urges Gorbachev to "Tear down this Wall"
		Gorbachev continues to restructure the Soviet Union, urging other Soviet bloc countries to follow his lead

YEAR	FAMILY	HISTORY
1988	In Switzerland, Cordula's teammate defects	
	Roland dies	
1989	Life goes on as usual for family in the East	Gorbachev influences fall of East bloc. Honecker resists. Crowd calls for freedom.
	Cordula becomes last East German champion in point track race	February: Last person killed trying to cross the Wall
		August: Hungary opens borders
		September: Demonstrators in East Germany: "We want out!" Police crack down.
		October: Honecker demands East Germany's fortieth anniversary festivities go on as scheduled
		Leipzig demonstrations
		November 4: One million East Germans attend prodemocracy demonstrations in East Berlin
November 9, 1989	Cordula trains in East Berlin	Berlin Wall falls. East Germans are free.
	Heidi in Karl Marx City	
	Hanna and Nina in Washington, D.C.	

PREFACE

The [Berlin] Wall is . . . an offense not only against
history but an offense against humanity, separating
families, dividing husbands and wives and brothers
and sisters, and dividing a people who wish to be
joined together.

—*President John F. Kennedy*

I was five years old when I learned that my grandmother lived
behind a curtain. The year was 1966. It was Grandparents Day in
kindergarten. In a slow-moving stream of little children holding
hands with the elderly, my classmates brought in their grandparents,
sweet-looking old people with pates of silky white or graying hair,
weathered faces etched in soft creases, twinkly eyes, and benevolent
smiles.

I sat at my desk, watching them come in. They greeted the teacher
cordially and one another as they shuffled in and made their way
into seats set up next to each child's desk. One at a time, my friends
excitedly led their grandparents to the front of the classroom and
proudly presented them to the rest of us, introducing them by names
like Nana, Poppa, Mimi—which were as foreign to me as they were

intoxicating—as their grandparents stood by beaming down at them lovingly. I was entranced by it all. Suddenly I felt alone and left out. I looked at them, then panned to the empty chair next to my desk, which got me wondering, where were *my* grandparents?

I came home from school that day wanting answers. I bounded through the front door, found my mother in the kitchen, and, without any greeting whatsoever, demanded, "Where are my grandparents?"

That evening, after dinner, my parents sat me down and told me why I had never met any of my relatives. Speaking in gentle but serious tones, my father, who was of German-Jewish background, explained that his family had all "died in the war." In my naïveté, I was unaffected, and turned to my mother, expecting to be disappointed by her as well, but was delighted to learn that her parents and family were alive. She brought out a photograph of her mother and said, "This is your Oma."

*O*ma. She was perfect. She looked exactly like the other grandmothers, but better. She radiated serenity and calm, a winsome, knowing smile gracing her face. Though I couldn't put it into words then, I was drawn to her pastoral elegance, her humility, her wise, confident disposition as she sat comfortably in an inviting, overstuffed polka-dotted armchair, her body positioned slightly askew, as she looked off to one side.

I stared at the picture for a long time, scanning her from head to toe, even cocked the picture just so, so it felt as if she were smiling directly at me. Despite the fact that I now know the photo was black-and-white, perhaps out of some subconscious desire to bring her instantly to life I saw Oma in color, pale blue eyes beneath those soft, heavy lids and what appeared to me to be blushing, pink cheeks. She had a simple upswept hairdo, the color the same brunette as mine, and wore two pieces of jewelry, a large brooch in the shape of a rose, probably made of gold, I thought, and a small pin that adorned a matronly black dress at the base of her V-neck collar. I imagined myself curling up on her plump, cushy lap and being swept up in an exquisite, warm embrace as she gazed down at me the very same way my classmates' grandparents had gazed down at them.

"Oma," I said aloud, charmed by the singsong ring of her name. Completely satisfied, I looked back up at my mother and asked, "When is she coming to visit?"

Unfortunately, my mother said, suddenly distracted as she got up to move about the kitchen, Oma could not visit us. Nor could we visit her. She was in a place called East Germany along with the rest of my mother's family, her sisters, brothers, and everyone else. I didn't understand, so my mother stopped, perched me on a kitchen stool, crouched down to meet me eye to eye, and explained.

When she was finished, I stared blankly back. Though I realize now that she must have used the term *Iron Curtain*, the only part of her explanation I understood at that moment was that they were in a place far away, trapped behind "a curtain." But this made no sense to me. I tried to comprehend why my mother would allow a sheer cotton panel like the kind I had on my bedroom window, or even heavy draperies like those hanging in our living room, to stand between her and her family. Someone, I thought, simply needed to pull that sheet of fabric to the side and let those poor people out. Someday, she reassured me, we might be able to meet them. Someday indeed. For goodness' sakes, I thought. It's just a curtain.

Nina, age five   Oma, behind the Iron Curtain

$\mathcal{I}$ went back to school the next day and told my teacher and friends that I too had grandparents, that my Oma was beautiful and, moreover, that I even had people called aunts, uncles, and cousins. My teacher was delighted. When she asked where they lived, I said East Germany, "behind a curtain." It was only when I saw her cheery face drop to a somber and sympathetic one that I realized the curtain might be bigger than I had imagined.

$\mathcal{I}$t would be several years into my childhood before I would discover that the Iron Curtain was not a simple panel of cloth that could easily be pushed aside by those on either side of it, but rather a symbol of something much bigger and more sinister than anything that childhood innocence could conceive. I would come to learn that my mother's entire family was indeed trapped inside East Germany and that my mother had escaped.

1985 | EAST BERLIN

In the musty East Berlin velodrome, with stopwatch in hand, the trainer blew the whistle and the East German women's national cycling team took off. Pedaling their single-gear track bikes, they moved easily, gradually accelerating as they made their way around the 250-meter pinewood track.

Winding their way around the oval in a graceful, measured cadence, they slid seamlessly into a tight paceline, their bikes only inches apart, their thin tires gripping the track's shiny lacquered finish.

Syncing technique with speed, they postured for position, increasing speed on the straights and banking on the curves. Then, when their trainers shouted for better form and more effort, East Germany's top-tier cyclists broke out of their tight formation and pushed at full bore.

By the next lap, they were full-on, jostling for control of the track, pushing with everything they had, their tires seemingly defying gravity, clinging to the track's slippery, steep-sloped walls by centrifugal force. They jetted down the straightaway, passing their trainers in a whoosh, the trainers yelling *"Weiter! Schneller!"* (Faster!) and the athletes responded, pouring themselves into every pedal stroke with ferocity.

Pushing, stretching, pedaling as fast as they could go, gunning *à bloc* as they gapped, dropping back an inch, then surging forward, standing out of the saddle, they forced the pace, trying to control their bikes, finally exploding in one last burst of speed until the trainers clocked it out as the athletes streamed in, one after the other, through the finish.

*J*ust a few miles down the road, believing we had not been followed, our intelligence team dropped off the highway in our olive-drab Ford and made our way into the East German forest on a dirt path that had been carefully chosen to conceal our movement in order to reach our target unobserved. We moved deeper into the silent woods, trying to avoid ruts as we drove carefully along the bumpy path, all the while scanning the wood line for any sign of danger.

Then, just as we began to move into position, a single Soviet soldier, weapon raised, stepped directly into the path of our oncoming car. Other soldiers appeared out of nowhere, immediately taking up positions around the car, cutting off any chance of escape.

With soldiers now blocking the path in front and rear, a Soviet officer made his way to the passenger side of the car, brandishing a pistol as he chambered a round.

The muzzle of his loaded pistol tap-tap-tapped against the glass. He ordered, "*Atkroy okno.*" (Roll down the window.)

When there was no response, the muzzle of the gun now fixed against the glass, he snapped, "*Seychass!*" (Now!)

PART ONE

Schwaneberg schoolhouse and church

1

THE HANDOVER
END OF WAR
(1945)

A mother's love knows no bounds.
—*Author unknown*

*O*ur story started when one war ended and another began.

The day World War II ended, my grandmother, Oma, was one of the
first in the village to emerge from the underground cellar and step out
into the still and desolate landscape of rural Schwaneberg. At forty
years old, her belly swollen with her seventh child, she hoisted open
the heavy wooden door and climbed up onto the dry, dusty landing
as her children followed, squinting as their eyes met the daylight.

Other village women and children emerged from the cellars of
their own homes, wandering about and awakening to what promised
to be a new day in Germany. With no able-bodied men around to
assist her, Oma directed her children to help her pull up the bedding
from down below where they had lived during the last two months

of the war, and move it all back upstairs into the living space of the family's wing of the schoolhouse. There would be no more overhead Allied bombing runs en route to their targets in the nearby industrial city of Magdeburg. Germany had been defeated, Europe had been liberated, and the skies were finally quiet.

It didn't take long for the village women to meet one another over picket fences to speculate when their husbands and sons would return. They wondered about what lay ahead for Germany and, most important to them, what was in store for their village of some 900 inhabitants.

Oma saw no use in dwelling on worry and set herself instead to getting her house back in order. Though school had been closed for months, she insisted her children return to their studies and get back to their chores, cleaning out the schoolhouse and scrubbing down the desks to prepare for a new school year. With food stocks all but used up and the ground fallow, the once-green potato beds emptied and parched to hard-cracked dirt, she directed the younger children to gather dandelion and nettle greens in the meadows and comb the berry bushes for any remaining fruit while the older children helped her prepare the soil for all the planting they had missed that spring.

When most of the men had still not returned after several weeks, a pall descended on the village. After only a few men came back, Oma began to wonder when, or even if, her husband and son were ever coming home. Opa, my grandfather, a forty-five-year-old schoolteacher and headmaster, and their oldest son, Roland, not even eighteen years old, had been pressed into service in the waning days of the war, when the Third Reich had ordered that every last able-bodied male over the age of fifteen join in the fight to the end for Germany.

*A*s the women waited for their men to return from the front, they became alarmed when stories seeped into the village that, as the Soviets were making their way onto German soil, they were raping and killing German women. Word spread quickly that Stalin openly encouraged rape and pillage as the spoils of war, a reward for Red Army soldiers for their sacrifices and the struggles they had endured against the German army, the Wehrmacht. Refugees passing through Schwaneberg on their way to the West confirmed the reports, recounting their own harrowing stories of savage assaults or telling of others murdered after a rape or when they fought their attackers. One family told a horrific story of their teenage daughter who was raped, then shot dead in broad daylight.

Women throughout Germany now feared for their lives. In Schwaneberg, they hoped that their men would return home in time to protect them should the Soviets enter their village. Oma became especially concerned for her oldest daughter, a pretty, wide-eyed, raven-haired seventeen-year-old—my mother, Hanna.

*B*y spring, American, British, and Soviet units were rolling into cities, towns, and villages throughout Germany to establish command and order. Oma, like most of the women in Schwaneberg, believing Hitler's denigration that the Russians were a barbaric lot, prayed the Americans or British would take their village. The American commander, General Dwight D. Eisenhower, some noted, even had a German name, which bolstered their hope that the Americans were more like them than the Russians were.

Then one quiet afternoon in mid-May, their wait came to an end. Everyone in the house and even the neighbors heard little Kai shriek from the loft upstairs. Her round, pregnant form slowing her, Oma made her way up the staircase of the east wing of the schoolhouse as her other children, Manni, Klemens, Tiele, and Hanna, bound past her. At the top of the landing, she found her little Kai surrounded by his brothers and sisters, pointing out the small oval window to some trucks in the distance. The family perched at the window in silence, anxiously waiting to catch a glimpse of which army was rolling in. The small convoy of three trucks inched closer, then stopped when it reached the edge of the village. Oma watched, her nerves in knots as she braced for a sign. From his window, the mayor unfurled a white sheet. The village mothers jumped to follow his lead, all of them, including Oma, hanging white sheets from their windows.

The trucks approached cautiously and finally came into full view. The children froze and Oma stared in disbelief until the older boys broke the silence with an ecstatic cheer. The first vehicle, marked with a white star, slowly led the convoy as it made its way down Adolph-Hitler-Strasse, and onto the cobblestone square. Down below, the mayor appeared from his house and quickly hobbled onto the street to welcome the Americans. Hanna looked at Oma, who smiled and gave her a nod, the go-ahead to take her siblings out to join the crowd quickly assembling outside.

The Americans stopped their trucks. From atop, they tossed Hershey's chocolate and gum to the village children, who were quickly disarmed by the soldiers' cheerful expressions and animated demeanor. As they passed out treats, the soldiers spoke in friendly tones with happy-sounding words that none of the villagers could actually understand. One soldier removed his helmet and hoisted Manni onto his jeep as other little boys looked on with envy. From up in their windows and down on the street, the village mothers watched the scene, waving to one another and raising their hands to the heavens in gratitude.

Over the next few days, nearly everyone became enamored of the American soldiers, their easy, open way, their childlike humor and lighthearted antics. For the first time in many months the women smiled, becoming particularly amused when the soldiers sent their children into fits of giggles when they botched German phrases, saying things like "Hello frowline. Itch leeba ditch," or calling everyone *Schatzi*, an endearment reserved for parents or for those in love.

Over the next weeks, the Americans established calm and control. They clowned around a lot, laughing, taking pictures of themselves with some of the village children, even assembling to get a group photograph in front of the Adolf-Hitler-Strasse sign, which one of them removed afterward to take home as a souvenir.

For the most part, though they endeared themselves to the community, a few vented their anger against the Nazis by taking it out on the villagers, looting and destroying personal property. Oma came home one day to find that the lock on Opa's desk had been pried open, the contents—a silver letter opener and heirloom box—stolen, and a swastika carved into the seat of his big leather chair. On the seat, like a calling card, lay an American penny. By and large, however, the villagers' worries began to subside, most of them believing that, under the Americans, their lives would get better.

*B*ut their relief did not last long. One day the Americans shocked the villagers with an announcement that they were leaving.

"Germany has been divided into two separate areas of administration," said the senior sergeant. The Americans and British would take command of the western part of Germany and the Soviets, the

East. Looking over the crowd, he said simply, "Schwaneberg will fall under Russian control."

The villagers were stunned. It was as if the bomb they had feared during the war had finally exploded in the village.

"There is nothing to fear," the sergeant continued, assuring them. "The war is over and the Russians will come not as fighting troops, but as a peaceful occupation force."

The crowd became agitated. Someone muttered about fleeing before the Soviets arrived. Hanna turned to Oma and suggested the family pack up and go, but Oma dismissed the idea. It was not a reasonable plan in the first days after a war for a pregnant woman with so many young children to flee, without food or shelter, without men for protection, facing chaos and uncertainty on the road with thousands of other refugees disappearing to places unknown.

"And besides," she said, "how awful if Papa and Roland came home to find that we had abandoned them."

Oma's attention was drawn back to the sergeant, who concluded with one last announcement.

"Should anyone have a compelling reason to leave," he said, "we are authorized to take a few villagers with us to the West." The women now looked at one another. Some fidgeted, some stepped back, most shook their heads, not willing to consider breaking up their families. They returned to their homes trying to console themselves about what the future might hold under Soviet occupation.

That night marked one of the most difficult decisions Oma would ever have to make. Sometime after midnight, she went into Hanna's bedroom and quietly sat on the edge of her bed. As she watched her daughter sleep, she studied Hanna's face and reminisced about her childhood, taking stock of her life, starting with the very night of her birth.

Hanna had come into the world on a bitterly cold, dark winter night in Trabitz, a tiny hamlet on the Saale River. Outside the schoolhouse that night the winds had kicked up enormous snowflakes that had wildly flown about all evening long and never seemed to settle. The rooftops and trees had been covered in a thick white blanket of snow and the rooms inside were ice cold. In the wooden-slatted garret of the one-room schoolhouse, Oma, not much older than her oldest daughter was now, prepared to give birth alone. Opa, a schoolteacher in his mid-twenties, had run off into the night to try to find the doctor. Their firstborn, ten-month-old Roland, slept soundly in a wooden cradle a few feet away. Then, in the stillness of the night, the new baby came into the world, its cries echoing throughout the hollow room. Oma cleaned the baby with her blanket and looked to see that she had given birth to a girl. Holding her newborn against her skin, she calmed the baby, who settled easily into the soft folds of her exhausted body.

As a little girl, Hanna had wanted to grow up quickly. While Roland had blossomed into an ideal child, his precocious little sister had been difficult. Roland was a parent's dream: obedient, smart, a natural-born leader. Hanna, however, the little curly-haired firecracker with steely blue eyes that constantly scanned the scene for any sign of adventure and mischief, was playful and impish, a rabble-rouser with endless energy who made her own rules, falling in line with her father's discipline only when it suited her.

By the time Hanna was four, Opa wondered why, unlike other children, she could not manage to sit still. Too young for school, Oma would hand her a hoe and set her to help in the garden. When

Roland, age three, piloting, with Hanna, age two, in 1929

Hanna grew bored with that, Oma put her in charge of feeding the rabbits grass and hay, which often ended with Hanna purposely failing to properly close the gates, then gleefully chasing the rabbits until they were caught and accounted for. In an effort to calm her restless spirit, Oma taught Hanna to knit but she quickly lost interest in that as well and asked Opa to teach her to read. At five years old she could read the newspaper. Opa occasionally brought her along to his Saturday-night card games at the pub and made her read words aloud like *Nationalsozialistische Bewegung* (national socialist movement) and *Demokratisierung* (democratization), enjoying it when his friends laughed in disbelief. Wanting badly to go to school, every day she sat outside Opa's classroom window crying, then wailing until Opa emerged from the schoolhouse and chased her away, ordering her to go home. At home she would cry some more but was always back the next morning outside the window to repeat the scene. At

(Clockwise from upper left:) Klemens; Hanna, age ten; Tiele; Manni, in 1937

her wits' end, Oma pleaded with Opa to take Hanna into the class-room and let her sit in the back. Unable to see her father from the back row, where she sat with the oldest and tallest children in the one-room schoolhouse, she remained quiet as a little mouse, stealing glances at her older, endlessly fascinating neighbors. After that, Oma bought her a slate board and tied a sponge and a drying cloth to the hole in the wooden frame and Hanna happily carried the board with her wherever she went, practicing letters and writing words whenever she saw them.

When Opa was promoted to become the headmaster at a bigger school, the family moved to the larger village of Schwaneberg (the Village of the Swan) in the district of Schönebeck in Saxony-Anhalt.

A storybook country village with stone buildings and half-timbered houses with red clay–gabled roofs surrounded by farmland, Schwaneberg had its own fresh bakery, a small dairy, two churches, a schoolhouse, barbershop, and horse stables, which surrounded the cobblestone square and dotted the main street. Mostly self-sufficient, it was supplemented by fresh vegetables and dairy from local farms, cheese from the Jewish door-to-door vendor, and pots, pans, and glittering trinkets sold by colorful, exotic Romany who traveled through the village twice a year.

Once the family had moved into the teacher's quarters in the east wing of the now two-story schoolhouse, Oma had decided that Hanna needed to be prepared for life and taught her what young village girls needed to know: how to plant and harvest a garden, to sew, to help with household chores and tend to younger siblings. But Hanna detested domestic work and often disappeared in the middle of a task; she would later be found hiding under a table with a book or running around outdoors, taking on the boys in street races or scaling the massive stone wall that separated the farm fields from the backside of the schoolhouse.

*O*ma came from a long line of farmers, proud salt-of-the-earth folk who had only ever known the labor of the land. Opa, on the other hand, hailed from an academic family and was the most educated man in the village. He played the violin, the harmonica, and the organ, but most of all he loved playing his Schimmel piano, a family heirloom he had been gifted by his parents upon graduation from teachers' college. He had insisted each of his children play an instrument, if not the piano then the recorder, the flute, the accordion, or the mandolin, all of which he kept in his office, propped up against his books on the shelves. He also taught his children to sing in harmony and showed off his singing troupe to anyone who would listen.

Over the years, Opa had amassed a large private collection of books. Dozens of tomes on art, history, geography, astronomy, wild-life, religion, science, and foreign lands and cultures, some in French and Latin, lined the walls of the family home. Inspired by his interest in the world beyond their village, his children and students alike learned about the Louvre in Paris, the Prado in Madrid, the Pinakothek in Munich, and the Hermitage in Leningrad. Hanna read books about America, becoming especially fascinated with Native Americans.

One day, when she was about eleven years old, she discovered a book in Opa's study about a famous fortress in western Germany. The magnificent Heidelberg Castle, nestled into the Odenwald Mountains high above the deep, green Neckar Valley, had once been one of the most opulent palaces of the European Renaissance. Built in the thirteenth century, the great castle had been home to powerful kings and other preeminent German royalty of the day. Hanna and her siblings instantly became infatuated with the fortress, imagining lots of exciting details about what lay hidden in its endless great hallways and dark dungeons, dreaming up stories of brave, sword-wielding knights rescuing princesses from winged, fire-breathing dragons, magical fantasies that no child could resist.

Not long thereafter, the children were thrilled when their father brought home a kit model of the Heidelberg Castle. Opa and his children spent every afternoon over the next few months meticulously piecing together corrugated boxboard parts, constructing the castle higher and bigger as they tended to every detail in its elaborate ornamentation, the precise emplacement of stone columns and gate porticos, properly positioning battlements and turrets, the drawbridge and even the iron ring door knocker, as their father regaled them with a captivating history of the city of Heidelberg. When the model was finally finished, Opa praised his children, then stood back and called it "a masterpiece." Some of the younger children, still believing the castle just a make-believe fantasy, were surprised when Opa assured them it was not a fairy tale, that the castle actually existed, and that there were many such marvels throughout the world.

"The world is infinitely vast and full of wonder," he had said. Then, paraphrasing Mark Twain, his favorite American author, he told them to "träumen, entdecken, erforschen"—explore, dream, and discover it. That day Hanna came to view the Heidelberg Castle as a symbol of the extraordinary world that lay beyond the lovely but ordinary provincial village of Schwaneberg.

Oma gazed down at Hanna, who continued to sleep deeply and soundly, and then she left the room.

Before sunrise, Oma got up, went to the kitchen, and filled a small burlap sack with a sweater, socks, and a few potatoes. Then she sat and waited.

Just before daybreak, when she heard the Americans starting up their trucks, she woke Hanna, told her to get dressed, and took her outside in the dark to the cobblestone square. A few villagers had already gathered to bid the Americans farewell. Then, in an instant, before Hanna even knew what was happening, Oma pressed the sack into her arms and, with a shove, turned her around and presented her to the American sergeant in charge. Jarred wide awake, Hanna realized what was happening and broke free, but Oma caught hold of her and pushed her decisively into the arms of the sergeant. Stunned, Hanna turned to reach back for Oma, but Oma took a few steps backward and stood resolute.

Soldiers atop the truck quickly pulled Hanna up and wedged her

Hanna, age seventeen

in between two others seated in the truck bay. She sat frozen, silent, staring down at Oma, but when the trucks began to move, she cried out. Oma did not react and said nothing as she held her daughter's gaze, trying to appear strong.

The Americans drove away slowly, their tires kicking up a din of dirt and gravel. Jostling with the movement of the departing truck, Hanna looked back through the dusty swirl to her mother, who stood rooted in place in front of the other villagers. As the convoy rumbled out of the village, tears rolled down her face as she watched Oma, her figure diminishing with the growing distance. The other villagers remained standing there, astounded at what they had just witnessed.

As the sun rose that morning, casting an amber hue over Schwaneberg, Oma watched Hanna disappear. She prayed that she had made the right decision and asked God to give her strength. Then she turned around and walked home without looking back.

*T*he convoy continued on its way, reaching the next village and picking up more army vehicles to join in their retreat to the west. In the truck bay, the soldiers tried to console Hanna, telling her that everything was going to be all right, but despondent, she sat silent, finally burying her head in the burlap sack.

Her mind wandered until it settled on the first time she almost lost Oma. She had been six years old. Oma was well into her last month of pregnancy with her fifth child. Hanna and her brothers had come home from school to find Oma collapsed on the kitchen floor. Opa had come bounding in with the doctor, shooed the children away, and hoisted Oma up and onto their bed. Minutes later the doctor emerged, telling the children to go into the bedroom and say good-bye to their dying mother. She was pale and seemed barely alive. The baby did not survive, but fortunately Oma did.

Several miles down the road, Hanna's thoughts began to race. Oma had given her a gift, a chance that would not come again, to safely escape life under the Soviets. Hanna envisioned Oma's face. Suddenly she panicked. Without warning, she catapulted herself over the side of the moving truck and hit the ground with a thud. The soldiers scrambled, shouting to the driver to stop the truck, but by the time the vehicle came to a halt, Hanna was already up and running home. Several hours later, she walked through the front door.

2

AN IRON CURTAIN DESCENDS
COLD WAR BEGINS
(1945–1946)

From Stettin in the Baltic to Trieste in the Adriatic, an
iron curtain has descended across the Continent.
—*British prime minister Winston Churchill*

By the time I arrived in Berlin to work for U.S. Army intelligence in the 1980s, East Germany was well established as a hard-line communist state. Because of its location on the Warsaw Pact's westernmost frontier, the Soviets had amassed a force of some twenty divisions inside the country, making it one of the most militarized places on earth. The Red Army sent its most lethal forces to East Germany to stand face-to-face against its NATO (North Atlantic Treaty Organization) enemies, the United States and her Western allies, located just over the border in West Germany.

But in 1945, East Germany was in its infancy. Known in its early years as the Soviet Zone, the nascent state had yet to be defined and so the Red Army descended on the eastern territory with a plan to

reshape the face of the East. The first challenge the Soviets faced was
to change the mind-set of the almost 19 million German citizens
who, long before World War II, had been led to believe that communism was the greatest threat to the Western world. Stalin demanded
the transition be swift, and the approach uncompromising.

The first hours of occupation would set the tone for the birth of a
new nation.

In the early-morning hours before dawn on the second of July, the
Soviets announced their arrival through a bullhorn, in a tinny echo
that startled the villagers awake. Speaking German with a heavy
Russian accent, a voice boomed over and over again, "*Achtung, achtung, achtung* . . . The Soviet Army comes in peace."

Russian soldiers in field uniforms dismounted the vehicles and set
to work establishing a small command post in the village square.

At daybreak, many of the villagers were peering out of their
windows, nervously watching the Russians from the safety of their
homes. Curiously the mayor was nowhere to be seen, but this time
the villagers did not need their mayor to tell them what to do. They
already understood by instinct to stay indoors until they were instructed to do otherwise.

Several hours later, another bullhorn squelch sounded and the
voice continued.

"The Soviet Army comes as friends and brothers to help build a
new Germany."

He informed the villagers that, in the coming days, there would
be many new laws and important changes. He imposed a curfew, requiring the villagers to be in their homes from nine at night until six
in the morning. A slew of frightening directives followed:

All food is to be relinquished to the Soviet Command immediately.

Anyone found hoarding food for himself or his family will be shot.

Anyone who attacks a Soviet soldier will be shot.

Anyone resisting or disobeying any law, order, or regulation set forth from this day onward, will be severely punished.

"Under orders from Stalin," the voice concluded, "any Soviet soldier who causes violence against German women will face serious charges. So," he concluded, "there is no reason to fear us." It was in essence a promise that Schwaneberg's women would be left alone so the Soviets could get on with the business of taking control and the villagers could focus on adjusting to change.

Most villagers dared not venture from their homes in those first days, but by mid-July they were beginning to accept their fate and emerged to face their new lives. They began to go about their business, carefully avoiding the Soviets as much as possible. The Soviets kept to themselves, not mingling with the Germans unless they were giving orders. They directed people to check the bulletin board at the local *Gasthaus* (tavern) to keep abreast of all new orders.

*T*he change in the village was jarring. The entire village cleared out what was left of their wartime stores, bringing forth their food to the drop-off point, which was staffed by armed guards: Oma sent her children into the cellar to collect and turn in their last stocks of potatoes and jars of pickled vegetables, items that had sustained the family through the final days of the war. At the depot, the Soviets promised to redistribute the food equally to the villagers, which they were slow or altogether failed to do.

Then the Soviets installed a new mayor. The man who had held the job for as long as anyone could remember was pushed aside. Herr Boch, the seventy-year-old barber, a man who had kept his communist leanings to himself throughout the war, became the mayor and mouthpiece for the Soviets. He strutted around the village showing off his new status, proudly wearing a red shirt his wife had sewn from an old Nazi flag.

Those first inchoate weeks under Soviet occupation saw a rapid ideological transformation in the East Zone. In Schwaneberg, Mayor Boch's zeal for communism took the villagers by surprise. He showed up everywhere, talking up the virtues of Marx and Engels, trying to invigorate the community to come together "for the greater common good," and promising a new future after Hitler had miserably failed them.

In Washington, the United States started to develop plans to improve conditions in the West Zone. In the East, as reparations to the victors, the Soviets stripped the land of everything they could cart away and send back to the Soviet Union. Entire cities and towns were gutted of anything of value, including industrial equipment, farming machinery, tools, building materials, furniture, bathtubs, even toilets, and hardware accessories as well, such as light fixtures and doorknobs. Railroad tracks and whole factories were dismantled and transported eastward to be reassembled in the Soviet Union. As the East was being plundered of its resources, people began to walk westward.

*U*krainian and Polish forced laborers who worked the farms for the Germans during the war were released to go home and Schwaneberg's women and children were made to take their place. Along with other villagers, Hanna was put to work in the fields.

One afternoon, after a long day of laboring in the carrot field, having gone several days with little food, Hanna became dizzy. Tempted to eat a single carrot, she looked around and, seeing only other village workers, decided against it for fear that the authorities would somehow find out and she would be shot. Despite Mayor Boch's assurances to the villagers that things would soon get better, when their children started passing out from hunger, the village women began to secretly pocket small bits of food and vegetables from the fields, taking their chances on being shot.

By late summer, when there was still no sign of Opa or Roland, Oma, now in her seventh month of pregnancy, knew she had to find a way to get food for herself and her children. She put Hanna in charge of watching over her siblings and traveled by train to her parents' farmstead in Seebenau, a small, rural community on the edge of the East–West border line.

There, Oma found her parents, Kallehn and Ama Marit, struggling to adapt to the new changes that the regime was imposing on farmers and their crops. While Oma worried about the toll the harsh new laws would have on her aging parents, Kallehn worried about his daughter and her many children.

Upon her arrival, he consoled her with a grin and a gleam in his eye as he showed her his "forbidden stash" he kept hidden under the cold-store floorboard in the storage house. Soon Oma was on her way home with a pack of treasures including a block of butter, a goose, sugar, and flour, thinking about the sumptuous roast with gravy she would make that would delight and reinvigorate the family. But her luck was not to last. At the train station in Salzwedel, a guard confronted her. Unapologetic even after noting her pregnant form, he simply confiscated her packages, shaking his head while condemning her for "depriving fellow citizens of essential food." Eventually she was released and arrived home with empty hands.

*B*y now Oma had made up her mind about the Soviets. She watched, disguising her worries as they stripped everyone of the things they owned, including their personal dignity. She was alarmed when she saw those who resisted Soviet authority taken away. Inasmuch as she feared the new regime, she knew they would all have to adjust.

Her teenage daughter, however, naïvely unafraid and unable to quiet the storm that brewed inside her, became irate and her anger boiled over when her mother came home without the food. With the other children, pale and weakened, looking on, Hanna stormed about the kitchen, railing about how anyone could take food from a pregnant woman with so many hungry children or shoot someone for eating a carrot. Oma hushed her, telling her to keep her emotions in check, warning that her rebellious behavior would get them all into trouble and lead to grave consequences for the whole family.

"We don't need to attract any negative attention," she reminded her children. "There's nothing we can do, so let's not complain or have any more speeches."

Despite her mother's call for calm, the idea had already started to take root in Hanna's mind that the new communist system was not something that she planned to be a part of.

Oma's words of caution to her children not to attract attention were indeed wise. Just days later, in a neighboring village, a teenage girl and her cousin were picked up by Soviet soldiers as they sat by the roadside in a lethargic stupor at the entrance to their village, asking for food. Weak from hunger, they had propped up a homemade poster made from a battered piece of cardboard, which read, "Dear Communist Party of Germany. Please give us food." This was clearly not the image the new authorities wanted to project, and the Soviets removed the girls from the street and carted them off to prison.

Throughout the East Zone, from big cities to country hamlets, the Soviets took control drafting German citizens to help in the administration of the new order. As near famine spread in the East Zone, the Soviets and their new German helpers established firm control and imposed new authoritarian governance.

While rapes continued in other parts of the Soviet Zone, owing to the strict orders of the Soviet commander in Magdeburg, none of the women in Schwaneberg was believed to have been raped. In fact, the Soviets kept their distance from the Schwaneberg population, establishing a clearly defined occupier-occupied relationship and conducting themselves in a strict, businesslike manner.

For the most part, the Soviets sowed fear in the villagers, but there were exceptions. Lieutenant Ivanov, a local commander, always behaved honorably, even gentlemanly, to the Germans he came across, occasionally giving them rides in his horse-drawn carriage and sometimes even giving them food, all without expecting favors in return. But the cost of real fraternization was harsh for both parties involved, as in the case of one sixteen-year-old girl who was arrested along with her mother. Both were given sentences of twenty-five years hard labor after the girl became pregnant by a Soviet soldier. The soldier was executed, the child taken from its mother, who was imprisoned.

In September, Roland walked back into the village.

His unit had surrendered to Red Army troops, and he spent nearly a month in a Soviet prison camp. There, the commander ordered his German prisoners to count off by tens. Every tenth man was released to return home to help rebuild the East Zone, the other 90 percent set on the long path to join the thousands of German POWs already making their way to the Soviet Union to serve as forced laborers.

Oma saw that the horrors of war had changed Roland, but he had matured and she believed he would become the new man of the house.

But a week later, just days before Oma gave birth, Opa too came home, having been released from Bad Kreuznach, a notoriously harsh American internment camp where, given the lack of food and exposure to the elements, hundreds if not thousands of German prisoners died. He arrived in Schwaneberg haggard and dazed. Having listened to Nazi propaganda demonizing the Soviets for over a decade, he was troubled to find his family living under Red Army control.

Despite the apprehension that had taken root in Schwaneberg, Oma was relieved to have her husband and son home. The family was fortunate: seventy of the village's men and boys never came back. With her family once whole again, Oma poured herself into guiding them to lead as normal a life as possible. With the assistance of the village midwife, she gave birth to her seventh child at home, a little girl, Helga. Though Opa could not help but think the timing for another child inopportune, the birth brought new optimism and he took the baby to show it to the rest of the children, who ogled over their new little sister.

Before long, the family began making preparations for a traditional christening, the baptism to be held on a Sunday in the village church, the way the community had always celebrated every birth. But that idea was dashed when Mayor Boch showed up to inform Opa that the building formerly known as the church was no longer available for such ceremonies, adding, "Communism is our religion now."

It was a shock to the family but, still hoping to get his job as a teacher back, Opa quietly accepted the edict. Oma kissed Helga on her little forehead and said a silent prayer for her infant daughter, the first of her children to be denied a proper church christening.

*I*n any small German village of that era, besides the mayor, teachers were the most revered figures, a status that Opa was proud of and took very seriously. At six feet four inches tall and large-framed with ramrod posture and penetrating gray eyes, Opa cut an imposing figure. His commanding presence, coupled with his ability to organize and unite the villagers, made him a natural leader. Ten years after he first arrived in Schwaneberg, the villagers regarded him as their de facto leader. One of the most influential people in the village, he was also the tallest, traits that did not go unnoticed by the Soviets.

Opa defined himself by his profession. He took his role as teacher very seriously, believing it a sacred calling to shape the youth of the village to be enlightened citizens of the world. He was also a strict disciplinarian who held himself, his family, and his students to a high moral code. He despised lying and cheating and harshly punished his students and children for such lapses. The community greatly respected him, even turning to him on matters beyond education,

Opa was ordered to teach a new curriculum based in Soviet ideology.

asking for his assistance on community or even personal issues and helping to resolve village disputes.

Eager to mold the community, the communists weeded out teachers they deemed "damaged by Nazi thinking." The rest were put through a "de-Nazification" process to rid them of any lingering fascist thought and to set them on a new course to spread the word of communism. While the other teachers didn't pass the test, Opa did and was asked to return as headmaster of the local school. With the Soviets expecting big things from him, Opa returned to his school alone.

The regime then had to recruit new, "untainted" teachers. Roland, calm and levelheaded like his mother, greatly revered his father and aspired to follow in his footsteps. Choosing to look beyond the harshness of the Soviets and believing in the promise that a post-

East German high school classroom with poster of Soviet leader Joseph Stalin

Nazi Germany future might hold, he signed up to train to become a teacher.

When school reopened in Schwaneberg, the Soviets directed that communist ideology be taught right from the start. Opa was instructed to portray Germany's war years as the disastrous result of a failed cult of Hitler's leadership, abetted by capitalist greed, that had deceived its citizens and led Germany to ruin. Opa's students, most of whom had been in the Hitler Youth, were now being told to radically shift their belief system and embrace communism.

Pressured to learn a great deal in a short amount of time, every evening after supper, when he would have preferred to have read poetry or play the piano, Opa crammed to learn Soviet doctrine, Marxist-Leninist theory, and Soviet history, so that he could teach the concepts the following day, knowing there would be a Russian monitor sitting in the back of the classroom making sure he got it right. Responsible for teaching the Russian language as well, Opa could often be found in his study, learning elementary Russian with one of his children sitting on his lap, repeating every word he said: "Okno, stol, pozhaluysta . . ."

A steady drumbeat of propaganda hailed the new regime's potential, but the forecast under communism looked dismal, so it was no surprise that people were still migrating to the West in droves.

By mid-1946, the Soviets were forced to take steps to formally control the exodus by posting Soviet troops along the border and erecting watchtowers, especially at heavily trafficked areas along the perimeter. Undeterred, people simply found ways around the control points, often easily walking out through forests and open fields. The Soviets responded by building up the border areas, then instituted an interzone pass system that required people to apply to travel to the West Zone. Now, by Soviet law, the only legal way to transit out, the application process was long and drawn out, meant to stall people who were determined to leave. Few applications were approved for healthy, able-bodied young adults who the regime feared wanted a one-way ticket out, but the authorities readily granted one-way passage to the aging, disabled, and infirm who they saw as nothing more than a financial drain on society.

A fter Opa had spent nearly a year as a teacher under the new Soviet masters, the authorities in Schwaneberg seemed pleased with him. Every day he looked and sounded more the part of the model communist educator. But before long, the man whom Hanna had always known to be proud and confident became unnerved. Oma watched him struggle as he tried to remake himself and appear loyal to the Soviet cause for the sake of the family's well-being. Despite his concerns about the integrity of his new teachings, Opa promised Oma he would do whatever was necessary to keep his wife and, with the birth of yet another little girl, Tutti, now eight children safe and his family intact.

*O*ne evening, Hanna approached Opa in his study, telling him that they should try to leave.

"It's too late," he said, resting his book on his lap. "And anyway, how would we leave? Where would you suggest we go—such a big family like ours?" Hanna stared back blankly.

"It will be all right, you'll see," he said, picking up his book to resume his studies, his words accompanied by an unconvincing, forced smile. Disappointed that she did not have an ally in her father, Hanna remained standing in place with nothing more to say.

Just beyond Opa, on his work desk, stood the cardboard kit model of the Heidelberg Castle that, years before, he had lovingly built with his children, with the promise that someday they would be able to see it for themselves. The Heidelberg Castle, which now lay in the American Zone, had never seemed farther away. The model, which had been built as a family project to be a gateway to the larger world, now stood as a reminder of their entrapment. Disappointed, Hanna lowered her head.

*T*hroughout the East, Germans became leaders in their communities and in local government as mayors, teachers, border guards, security officials, and as policemen in the newly organized police force, the Volkspolizei, or VoPo, the People's Police. They signed up to serve the Socialist Unity Party, a communist political party with a Marxist-Leninist ideology. A Stalinist-style single-party dictatorship, it closely resembled the Communist Party of the Soviet Union.

From the start, the regime made it clear that opposition to the Communist Party was outlawed and would be persecuted, so, along with scores of others, including the majority of village men who had recently returned from the war, Opa joined the Communist Party.

*I*n Schwaneberg, the family's situation deteriorated. While the older children somehow managed to get by, all too often when supper was finished, the younger ones, including four-year-old Kai, often lingered at the table, waiting for more. In particular, Kai missed the taste of milk, which was now so heavily rationed that it amounted to just enough to feed the village babies, so Oma shared Helga and Tutti's baby rations with Kai.

With so many little mouths to feed, in an effort to alleviate their hardship, Oma and Opa sent several of their children to live temporarily with other relatives. Hanna, now eighteen, was sent to live with her grandparents, Kallehn and Ama Marit, in Seebenau, in the Altmark district of Saxony-Anhalt, on the edge of the East–West border, and just a dash away from the British Zone.

3

"IF YOU WANT TO GET OUT, DO IT SOON"
CLOSE CALLS AND ESCAPES
(1946–1948)

Judge each day not by the harvest you reap
but by the seeds you plant.
—*Robert Louis Stevenson*

Kallehn sat waiting at the train station atop his rickety old farm wagon drawn by two horses. A serene old man with a shock of white hair and a weathered, whiskered face, his eyes lit up and he smiled when he saw his granddaughter. Delighted that she had arrived safely, he took her hand and helped her climb up to take a place beside him. As they rode, he did not talk much, but every so often turned his face to her, his eyes softening into slits, his whole face turning into a big grin.

Hanna settled in with Kallehn, whom, for some reason, no one knew why, everyone called only by his last name. He and Ama Marit, Hanna's grandmother, who preferred carrying the appellation of her

Kallehn (*standing at left*) in Seebenau on his farmstead

Old Norse ancestry, were happy to have Hanna's company. Hanna attended the local high school and helped Ama Marit with the household chores and helped Kallehn in the fields on the weekends.

Kallehn's farm had been in the family for almost two hundred years. A proud man and dedicated farmer like his father, grandfather, and great-grandfather before him, he had spent his life cultivating his land to provide the harvest that helped sustain his community. He was a deeply happy man whose greatest passion was to sink his hands into the rich earth and bring forth what nature had produced to supply life-giving nourishment. He worked hard and never complained. But now he faced the reality that one day soon, the regime would likely take away his land, would strip him of his livelihood, his heritage, and all that had ever mattered to him.

One day, as Hanna helped Kallehn and his farmhands harvest the fields, she discovered how being forced to hand over his crops had already begun to affect him. Though he hid his dismay in the creases of his weather-worn face, she could see that he was bothered that he would no longer be able to make his own decisions about his farm.

As the sun went down on that long day of hard work, Kallehn leaned back against a freshly packed bale of straw and sighed, "Today we have done our part for 'the People.'" With a smile and a wink, handing Hanna a small empty burlap bag, he said, "The gleanings are for us." Then he got down on all fours and, with his nicked and gnarled fingers, combed around the stalks, searching for tiny husks of leftover oats. As Hanna watched her grandfather hunched over and scrounging for the little bits on the ground, she wondered how long he would manage to keep that twinkle in his eye.

On Saturday afternoons, Hanna went to the local *Gasthaus*. It was not much more than an old building that had been turned into a place where young people could socialize before curfew, drawing classmates, farmhands, neighbors, even newly minted German border guards.

At the *Gasthaus*, Hanna met Sabine. Sabine had once lived and attended high school in what was now the East Zone, but, with postwar zoning changes, her house now fell in the British Zone. Always armed with the proper documentation to be sure she could get back home, Sabine occasionally crossed from West to East just to mingle with old school friends. Lucky her, Hanna thought, that she could move back and forth freely. The two quickly became friends and, before long, Sabine began smuggling in little presents for Hanna, passing them off when no one was looking: a piece of chocolate, a box of sugar, a pair of leggings, none of which were available in the

East and all of which were readily available on the burgeoning black market in the West.

*B*y now, at points all along the East Zone perimeter, wire fencing and guard posts were installed and roving patrols dispatched. With border guards now firing at will if they suspected a breach of the border under way and harsher prison sentences imposed for trying to get out without the proper paperwork, opportunities for crossing into the West diminished. On the Seebenau border, in light of the greater numbers of people fleeing, police presence increased and security tightened. More and more Germans signed on to augment the Soviet border guard force.

In Seebenau, the authorities kept a close eye on young people. Curfews were enforced, started well before sundown and ended after dawn. Punishments were now levied on not just those who attempted to flee, but also those who were suspected of having had knowledge of an escape and had failed to report it. Words against the regime were enough to have one escorted to the town Commandatura, the local makeshift Soviet headquarters, a converted stable, where they were interrogated and, afterward, often hauled off to prison.

It took some longer than others to get the message.

One day at school, as students were milling about at the back of Hanna's classroom during a break, Dieter, a likable but sometimes mischievous boy who sometimes had a tendency to talk too much, went a step too far.

"How can they teach us this slop," Dieter scoffed. "Can you believe it, teaching us that Stalin is 'the Great Leader.' Two years ago, the same teacher was teaching us that Stalin was 'the Great Demon,'" he said, putting his wiggling fingers to his head and sticking out his tongue in a silly mock-devil gesture. As the other boys guffawed and snickered, Dieter looked up to see the school's Communist Party minder staring at him with a penetrating glare. He approached the boy, seized him by the scruff, and hauled him out of class. Dieter was not seen again.

The day that Dieter disappeared, something in Hanna changed. She was alarmed the people seemed willing to accept the changes without fighting back. That very day, as she walked home from school on the path that led back to her grandparents' farmhouse, for the first time she took a close look at the border frontier. Scanning the horizon, she spotted a lone uniformed Soviet guard with a rifle who stood some distance away, smoking a cigarette, watching her. Not wanting to attract undue attention, she turned away and continued along the path the mile or so back to Kallehn's farm.

In the kitchen, Kallehn and Ama Marit listened to Hanna tell them about the boy who had been pulled from class.

"Where do you think they took him?" she asked.

"Maybe to prison," said Kallehn.

"What do you think about that, Kallehn?"

"What do I think?" he repeated. "I believe times are changing."

With that he simply went back to sipping his scavenged ersatz rye and oats coffee, his sun-chapped hands curled around the steaming cup filled with the mixture that he had made from his gleanings.

Now, in her most private thoughts, Hanna began to think more about, as Sabine had called it, "life on the other side." Telling no one, over the next weeks, Hanna probed the border, started to take note of guard shift changes, patrolling practices, trying to detect the differences between East German and Soviet border police procedures. Careful not to be noticed, she surveyed paths, tree lines, guard posts, paying particular attention to areas where she had heard someone had attempted escape.

In the East, Germans stepped up to take over positions in the government, signing on with security forces, staffing the newspapers, and administering schools, factories, and Communist Party and youth groups. In their efforts to show allegiance to the new ideology and to become catalysts for hard-line change, many approached their tasks with even more ferocity than the Soviets, cracking down ruthlessly on their own people. Scores of citizens applied to leave. Everyone knew the interzone pass was the ticket to freedom but they also knew the likelihood of being granted approval to transit was slim. As a result, hundreds of thousands took their chances trying to flee any way they could.

With the exception of Berlin, where travel between zones was still relatively unrestricted, essentially uncontrolled as a result of the Four Powers—the United States, Great Britain, France, and the Soviet Union—agreeing to administer the city together, security along the entire length of the border separating East and West Zones was tightened. In Seebenau, the border authorities rolled out barbed wire, erected more warning signs, and increased the presence of border guards. Only two years after the Soviet Zone had been established, detention centers and prisons were already processing prisoners for escaping, charged with what the regime called "trying to deprive the state of its labor force."

*O*ne evening at the supper table, Kallehn looked troubled. He rubbed his temples and told Hanna that someone had been murdered at the border on the south side of Seebenau.

"She tried to leave with a guide who said he could get her across," Kallehn sighed, "but this so-called guide took her money, her belongings, and then he killed her. Times are crazy, Hanna," he warned. "Don't go near the border. This is not a game."

Like a leaf blowing around in a wild storm, Hanna spent the next weeks anguishing over whether to succumb to a stifling life in the East or make a run for it. The day Kallehn showed her where the Soviets had dismantled railroad tracks that had once been connected to the rail line heading west, she decided the time had come.

She set out on a Monday afternoon after school. One of Kallehn's farmhands had told her about a border area just southwest of the rye fields, where guards paid little attention and, he said, even slept most of the time. Just to the north side of that post, he explained, one could easily cross.

She dropped off the dirt path and wound her way in toward the border. Once in the grassy pasture, she crouched, looking for any sign of activity. The border was quiet. The coast clear, she dropped her book bag, then moved to the edge of the forest, taking up a position behind a copse of pine trees. She panned the perimeter one last time and, seeing no one, slowly advanced. Careful not to snap a twig or rustle the leaves beneath her feet, she kept watch, her senses keenly sharpened to every possible movement and noise around her.

Suddenly a shout, firm and commanding, cut the air: "*Stoi! Strelyat' budu!*" (Stop or I'll shoot!)

But she didn't stop. She bolted like a deer who knows it is prey in the sight of a determined and steadied hunter. Carefully he tracked his target, a teenage girl, with long, dark braids flapping wildly on the back of her blue and white flowered dress, as she darted into the woods.

Pitching crisscross like a rabbit to avoid the sentry's aim, her breaths short, her heart pounding, she pushed past gnarled branches and bore through brush and briers that sliced her arms and scraped her legs.

"*Stoi!*" he shouted again, louder, emphatically.

Powered by fear and driven by the simple belief that she knew she could make it, she tore through the woods.

Then a crack. A bullet whizzed by her head and she dropped facedown into a ravine. Within seconds, a Soviet border guard was standing over her, his boots inches from her face, his rifle pointed down, directly at her head.

Kallehn was summoned and, after his profuse apologies, the guard released Hanna into Kallehn's custody with only a stern warning.

Now Kallehn worried constantly about his granddaughter. No longer willing to take the responsibility for her being killed at the border or banished to prison while in his care, he contemplated sending her inland, back home, to her parents in Schwaneberg. Hanna implored Kallehn to keep her escape attempt a secret from Oma and Opa, which he did.

Then one early winter evening, as she helped him heat bricks in the oven to warm the beds, he asked her what she wanted for her future.

She stopped and looked into his eyes.

"I want to be free, Kallehn," she said.

He looked at her sympathetically.

"What would *you* do?" she asked. "What would you really do if you were me?"

Kallehn dropped his gaze. He knew what he would do. Had he been younger, quicker on his feet, and not had the farm and Ama Marit to care for, he would have fled by now.

Looking up he said resolutely, "If you want to get out, do it soon." Then, reaching for another brick, he continued, "In less than a year, this place will be one big prison."

*T*he winter arrived, too cold a time of year to even consider escape. Hanna returned home to spend December with her family in Schwaneberg. With no church services to mark the Christmas holiday, Oma was determined to celebrate anyway and she and the girls decorated the house with red-ribbon-laced straw and wooden ornaments and bunted the mantel with fresh pine. She encouraged Opa to lead the children in singing carols just as they had done every Christmas, though this time they sang softer in case anyone might be listening. With her greatly reduced stores of flour, eggs, and sugar, she baked a modest version of her cinnamon, clove, and honey Pfeffernüsse cookies, managing to stretch her ingredients far enough for everyone to indulge.

Despite a bitterly cold winter and shortages of nearly everything the family needed, including coal to heat the house, laughter filled the family's wing of the schoolhouse, and after a while, no one even seemed to notice anymore that Christmas in eastern Germany was no longer welcome. As soft, crystalline snowflakes fell on the frigid environment that had become the East Zone, Oma, oblivious to the outside world, looked on with a smile, taking comfort in believing that the family would always be all right, no matter what the future might bring.

Opa continued to appear to adapt to his role, understanding well what the authorities expected of him, adequately teaching communist theory, lecturing about the dangers of American imperialism and that communism was Soviet Germany's true destiny and only salvation. He had worked hard to portray himself as the ideal communist, and so his Soviet and East German bosses rewarded him for his conformity and his mastery of Soviet teachings by recognizing his efforts at the village Party meeting. But at home Opa was not himself. He slept poorly and Oma began to bear the brunt of his bad moods.

Unintentionally, Hanna only fueled the fire. One afternoon, she confronted him.

"Papa, do you really believe in what you're teaching?" she asked. "Do you believe that there is truth in Marxist theory and communist teachings? And what about—"

"You're too young to understand," he cut her off.

"I'm not too young to understand," she replied. "I'm almost an adult."

There was no response.

"Then please explain it to me," she pressed.

He sighed but didn't answer.

"They don't really care about people," she continued. "They just want to control everything."

He remained silent, but she saw his temples pulsating, a sure sign that he was unnerved and on the brink of boiling over. She spotted the model of the Heidelberg Castle and tried again, this time more gently, and with a different approach.

"Papa," she said, dropping to his feet. Looking up, she pleaded, appealing to his sensibilities. "You were the one who said we should see the Heidelberg Castle," she said softly. "Get out into the world, you said. Explore, dream, discover."

"Times have changed," he said interrupting her, his voice growing louder, and then he snapped, "The Heidelberg Castle is now in the *West!*"

Hearing Opa's raised voice, Oma came scurrying in, shooed Hanna out, and tended warmly to Opa, patting his back and trying to calm him back down.

How much had changed, Hanna thought as she observed her father, since the Nazi days when she had her first boyfriend at the age of nine and her father had forbidden her to associate with him because he suspected the boy's father was a communist.

*A*fter the new year, Hanna returned to Kallehn's farmstead in Seebenau to finish her last semester of high school. Opa had informed her that, following graduation, she would begin training in a vocation for her future. That notion jolted her and she became alarmed at the idea of being trapped for the rest of her life in the Soviet Zone.

One afternoon as she and Kallehn shoveled snow from the walk, she stopped and said, "I've thought it through, Kallehn. I want to go." Kallehn too stopped shoveling. Resting his arm on the top of the shovel, he looked at her for a long time. Then he smiled—a sweet, sad smile, she thought. He dropped his head and sighed. Then he looked up again and said, "I will help you."

A few days later, Kallehn's other daughter, Oma's sister, Hanna's aunt Frieda, came to visit. That night, as Hanna slept, Kallehn discussed the issue with Frieda.

"She's young," Frieda sighed. "She has a right to choose her own destiny. If I wasn't so old, I'd have left long ago." With that, Kallehn and Frieda devised a plan for Hanna's escape.

*I*n early spring, Hanna pulled Sabine into the plan. Then, one warm, moonless night in May, only days before Hanna's high school graduation, when no one would suspect graduating students would be attempting to flee, Kallehn whisked Hanna off into the darkness. In his horse-drawn wagon, they traveled to Frieda's farmhouse in Hestedt, on the edge of Seebenau, near a stretch of border usually guarded only by roving patrols. There, in a dimly lit room in the middle of the night, a young man came into the house through the back door. Frieda paid him, Kallehn and Frieda hastened their farewells, holding back tears, and the young man took her off into the woods.

Silently, under the cover of complete darkness, they ran, swiftly slinking through the border and on into the West. Once safely on the other side, the young man deposited Hanna at Sabine's house, where the entire family was waiting for her with open arms.

The next day Kallehn, bracing for an angry reaction, sent word to Oma and Opa to tell them that Hanna had fled, though he made no mention of his and Frieda's roles in helping her to flee. Opa was livid and worried about the impact he believed her escape would have on his job and his ability to take care of the family. Oma just worried about what might have happened to Hanna.

In the West, Hanna, felt sheer relief. Thrilled to have made it out unscathed, she hailed Kallehn and Frieda for their assistance in facilitating her escape.

*S*everal weeks into Hanna's freedom, in the West, there was a knock at the door. Sabine's mother opened it and was startled to find a messenger from the East Zone with a letter.

He asked for Hanna and told her, "You must come back immediately. Your father is ill."

Sabine's mother came forward and yanked the letter from the young man, muttering that it was a ploy to coax Hanna back. She cursed, then shooed him away like a fly and started to shut the door. The messenger pleaded, asking Hanna to read the letter he said was from her mother.

In Oma's handwriting, Hanna read, "My dear Hanna. You have committed a crime by depriving the new free Germany of much-needed labor. If you come home voluntarily, there will be nothing to fear." Uncharacteristic of Oma's speech, Hanna detected a deception and shook her head in disbelief. The young man stood there until Sabine's mother pushed him off the stoop and slammed the door.

That evening, the West German police came to call. The police in the Soviet Zone had an arrest warrant for Hanna. In those early years, and surprisingly, given their vast differences in ideologies, a newly enacted agreement required the police forces of both zones to work together, assisting one another to enforce the laws of their respective territories. At twenty years old, Hanna was still underage, a minor, and therefore, in fact, in the West illegally. Though it was not a task they particularly enjoyed, it seemed the West German police were obligated by law to arrest her and return her to her parents in the East. Hanna refused to go. Sabine's father, a well-respected local West Zone government official, pushed his way to the door.

"Boys," he said smiling, "this has all been a big misunderstanding. I'll take care of it from here." With that, he ushered the policemen out the door. Though the police knew that Sabine's father had no legal authority in the matter, everyone just hoped the issue would simply fade away.

But with the Soviet police threatening Opa with the loss of his headmaster position as a first course of retribution, the issue did not fade away. A few nights later there was another knock at Sabine's door. Sabine opened the door.

"*Mutti?*" (Mama?) Hanna said, in disbelief, as she slowly rose from the sofa.

Speaking barely above a whisper, Oma told Hanna to get her things and come home. "Papa is very angry," she said. He had warned her not to come home without Hanna. Oma thanked Sabine's parents for caring for her daughter and declined an offer of a cup of tea. Dazed and conflicted but unable to deny her mother, and feeling ashamed at having put her in such a precarious position, Hanna packed her things and they went on their way.

Up the road, the two met up with Frieda, who had admitted her role in the escape and was now aiding in Hanna's retrieval to try to set things right with her sister and her infuriated brother-in-law. Oma and Frieda had sneaked out of the East without alerting border security and now the three had to make their way back in undetected.

With Frieda leading the way, Hanna, still stunned, in the middle, and Oma trailing, they set out for the border. Stumbling in the darkness, eventually they reached the barbed-wire perimeter. Hanna found an opening and stretched it wide for the two women to crawl through. They squeezed through, the barbs pinching and scraping, ripping Oma's dress. Once on the east side of the wire, Frieda then held it open for Hanna to cross through. With the two older women now firmly in the East and Hanna still standing in the West, all three stood looking at one another. Then, to Oma's great relief, Hanna climbed through.

Surrounded now by small clusters of trees, they would have to cross a large open field some distance from the Soviet guard post to reach the woods on the far side of the field before finally reaching Frieda's farmhouse. Frieda waited for a large cloud to move in front of the crescent moon and then moved, the others following closely behind. When the cloud passed, casting some light on the field, they fell cumbersomely into a flattened, prone position on the ground. Again they waited for a cloud and once again they rose and ran. Apart from the crunching of the brush beneath their feet, the pine forest was silent.

Suddenly, from across the field, they heard men's voices: Russians talking, laughing, as they walked along the border path. Then shouts: "*Stoi!*" (Stop!) The women scrambled, throwing their arms up. The guards approached, one noting that they were *zhenshini*, women, and lowered their guns.

In broken Russian, Frieda explained that they were returning from visiting a sick relative. One of the guards looked them over and told Oma and Frieda to continue on their way eastward, but said that Hanna would have to go with the soldiers. Oma linked her arm into Hanna's, declaring, "I go where my daughter goes."

The soldier moved to separate them, but Oma fought him off. He drove the butt of his rifle into her back and she fell to the ground with a thud, but didn't let go. He kicked her, trying to pry her loose, but she had grabbed Hanna's leg and was holding on tightly. Then Oma let out a bloodcurdling scream. The soldiers stopped short, looked at one another, finally cursing and waving all the women off to go home.

In Schwaneberg, Opa received Hanna with stony silence. Several hours later he called her to him, branding her a black sheep who had brought shame and embarrassment to the entire family and had endangered them all as well. He forbade her to have any communication with Kallehn and Frieda, whom he now completely distrusted. He ordered Hanna's brothers and sisters to keep their distance from her, in part to punish her but also to keep her from infecting them with her dangerous ideas. Their eldest sister of eight siblings, whom they had looked up to like a second mother, suddenly became an outcast in her own family.

*O*pa refused to let Hanna go anywhere alone. He organized things so she would be watched round the clock. It infuriated him no end that her reckless escapade just days before graduation had left her without a high school diploma. Oma and Opa now had to think about how to salvage her future. Perhaps Opa could pull some strings, he said, and she could still become a teacher. In the meantime, he set her to work in the fields for a local farmer, who was tasked, along with other villagers, to watch her every move and report to him if she did anything out of line.

At home, Hanna helped Oma with the chores and with the younger children. Oma found it hard to stay angry at her, but Opa remained cold. Not wanting to trigger his ever-growing anxiety, the children did not speak to Hanna in his presence, not even at suppertime. Hanna stewed, disappointed that Opa had not even tried to understand things from her perspective, as Kallehn so easily had.

After several months under his ever-watchful gaze, Opa allowed Hanna to resume her relationship with her siblings. She quickly reengaged with her brothers and sisters—running races in the grassy schoolyard's long fields with Klemens and Manni; having endless, sisterly talks with Tiele about books, and about boys, while they made daisy and cornflower crowns in the meadow. She tried to teach little Kai how to play the piano and took him for piggyback rides to the pond, where she taught him how to skip pebbles across the surface of the water. She helped Oma with the babies, giving them baths, dressing them, and putting bows in their hair. They had all missed their oldest sister terribly and were happy to finally have her back.

*I*t took nearly a year for Opa's anger to finally subside, during which time Hanna remained under virtual house arrest. By spring, he thought it best to get Hanna back on track, and so called her into his study to discuss her future.

He told her she would make an outstanding teacher. She listened as he talked compellingly about his life's work, his gratifying profession, how he enjoyed the intellectual challenges of being an educator and also the elevated status in the community that went along with the profession. Roland joined in, gushing about how proud he was to be an educator in a new era for Germany, helping to raise a new generation of postwar German citizens.

As he spoke, Hanna realized that so much had changed. Roland was adapting well to the new system whereas she was not. During the war years, they had been the closest of friends, daring to dream big when their future seemed so uncertain. He wanted to become a veterinarian. She wanted to be a lawyer. Somehow, they had decided, they were going to find a way to live and work together. They had made a pact then to stick together, no matter what may come, believing that, as a team, they could take on the world.

But everything had changed. Inasmuch as she loved and respected Roland, she simply could not understand how he could give in to the Soviets, how he could find pleasure in teaching communism and skewed theory, brainwashing unsuspecting children into believing lies and promoting a repressive society that functioned on fear and intimidation. But of course, Opa and Roland didn't see it that way. Like so many others in the East, they were hopeful, believing the system could work and that, with the family together, everything would be all right. She listened but said nothing.

"Hanna," said Roland, his eyes alight with dreamy enthusiasm. "This is a new beginning for Germany, don't you see it? We have a chance to really make our country great again. This is how I'm going to make a difference. You should do it, too." Hanna listened to it all and nodded, but when the discussion was over, she left the room, leaving Opa and Oma looking at each other and wondering.

From the start, the United States and the Soviets had radically different plans for Germany. To help Germany get back on her feet, the Western Allies, led by the United States, began to introduce reforms. The United States initiated the Marshall Plan, pouring billions of dollars into Western Europe to aid in recovery. To get the economy moving, the Allies helped Germany issue a new postwar currency. The Soviets were furious. They wanted to keep Germany fractured and weak and saw the U.S.-led plan as a threat. They ordered the Soviet Zone to continue using the prewar currency, which had become virtually worthless.

As a result, an economic wall descended between East and West. Though collaborators against a mutual enemy in war, the Soviets now made it clear that they had no intention of any further cooperation with the United States or its Allies.

Then in June 1948, in the first major crisis of the Cold War, the Soviets tested the Allies' resolve by severing all supply routes into the island city of West Berlin. Believing it critical to keep sovereign

territory defined by a Western democracy situated deep inside communist territory, the Allies stood up to the Soviets by launching a
massive aerial resupply effort.

Over the course of almost a year, in a near-impossible feat, some
200,000 flights provided food, fuel, medicine, and basic necessities to
the two million citizens of West Berlin. In the end, the Berlin Airlift
would prove a victory for democracy, putting a critical landmark at
center stage in a showdown that would pit Soviet authoritarianism
against Western freedom for the next forty years and, in no uncertain terms, define Berlin as the definitive front line of East–West tensions in the Cold War.

To Hanna, the Berlin Blockade, the currency division, and the cutting of the rail lines were major signs that the Soviets were committed
to permanently distancing themselves from the West and severing
the East Zone from the rest of Europe. This development meant only
one thing: Hanna had to escape for good, or she could be trapped in
the East for the rest of her life.

So in early July 1948, Hanna approached Opa and told him that
she indeed wished to become a teacher and asked to begin training
as soon as possible. Opa was ecstatic. He smiled, patting her on the
back, and told her she had made the right decision and would not
regret it. She looked back up at Opa, envisioning Kallehn in his
place, hearing his foretelling words: *In less than a year, this place will
be one big prison.*

4

FLIGHT

A SMALL SUITCASE AND
THE FINAL ESCAPE

(August 11, 1948)

None who have always been free can understand
the terrible fascinating power of the hope of freedom
to those who are not free.
—*Pearl S. Buck*

*O*pa arranged for Hanna to attend a daylong summer registration session in Magdeburg, some twelve miles to the north, before beginning teachers' college that fall. That day she was scheduled to leave in the morning and return later that evening.

The night before she left for Magdeburg, as she did every night, she helped Oma prepare the evening meal, on this evening paring and cutting up carrots and potatoes and sliding them into the boiling water. Unlike other evenings, however, she did not engage in the usual conversation about the day's events. Instead, she studied Oma as she moved about the kitchen: her small but robust form wrapped in her favorite faded blue-posy apron; her rosy, ruddy farmer's daughter's cheeks that gave her a perpetual glow; her ever-practical, matronly bun pinned on the top of her head. She moved purposefully about the kitchen, throwing herbs in the pot, making her way around the little ones who clamored to taste what she was cooking. Suddenly she looked up, catching Hanna's lingering gaze, and told her it was dinnertime. Hanna took off her apron, picked up the baby, and called out to the family, who came bounding to the dining room.

She tucked Tutti into her high chair at the table before sitting down at her assigned place as the oldest daughter, to the left next to Opa.

As they arrived in ones and twos, she began looking everyone over and taking in as many details as she could. The other children, thinking it was an ordinary supper on an ordinary Tuesday night, gathered around the long, pinewood dinner table, squabbling and prodding one another as children do, and tending to their dinnertime tasks: Tiele and Manni arranging place settings and drinking glasses, Kai setting the cutlery, Klemens helping little Helga onto her chair, and all of them, with the exception of Hanna, chattering, giggling, and joking with one another as they always did.

Roland came bounding into the house and went into the kitchen to give Oma a peck on the cheek before sitting down. Opa ambled in, greeting everyone, his large frame making its way to his seat at the head of the table.

From the kitchen, Oma emerged with a wide smile on her face, carrying in a stoneware terrine of steaming soup. Everyone reacted with glee, inhaling the aroma of sweet bouillon, looking longingly from their places in anticipation of consuming it. While Oma ladled the soup, Roland said something that made Klemens laugh,

Opa scolded Kai for coming to the table with dirty fingernails and, with Opa distracted, Manni tried to reach Tiele's feet with his own under the table without his father noticing. This was the way Hanna wanted to remember it: the smells, the noise, the activity, the laughter, the familiar chaos, the family together.

$\mathcal{A}$s supper began, Hanna looked around the table from one sibling to the next, at each of them in turn, trying to imprint the details of their faces and their personalities on her mind.

Roland, tall and handsome with a broad brow and movie star good looks—he beamed when he smiled, which affected everyone who ever saw him. A man now, he was a teacher, living on his own, home only occasionally for a home-cooked meal. Strong of character, with endless stores of positive energy, he was a true idealist always searching for truth, and a dynamic leader who saw the best in everyone. Hanna knew she would miss him, her closest sibling, terribly but had to believe that he would find his way under the communists.

Klemens, lean and athletic with doleful eyes, had a gentle, haimish way about him; the quiet, introspective brother who never craved the spotlight. Klemens had grown tall like Opa, but didn't sit as ramrod straight despite his father's constant prodding, and was much more reserved.

Sweet, sweet, little baby-faced Kai. Now six years old, with his silky blond hair and little button nose, he constantly amused his siblings even now as he mimicked Klemens's every action, when he put his elbows on the table, when he took them off, how he gestured with his hands when he talked.

Tiele, nearly a teenager, mother's little helper, loved to sew and bake and take care of the babies. With a scrap-fabric homemade bow in her brown curls, she wore a butterfly appliquéd cotton sweater she had knitted mostly by herself. She could spend endless days frolicking in the fields, making things from flowers for the little ones and for Oma.

Though barely two years old, the baby, Tutti, with her large, wideset brown eyes and wild, curly blond locks, was already a force, boisterous and sassy, and could never sit still. She lolloped in her high chair until she was given a piece of bread to keep her still. At three, Helga was quiet, observant, and sensitive. And Manni, good-natured and jovial, found humor in absolutely everything, even now as he made funny faces trying to get the others to laugh, until Opa told him to stop it.

Opa called the room to attention, saying with a smile, "So Hanna has a big day tomorrow." From across the table, Roland chimed in, "Hanna, you *certainly* have a big day tomorrow. We are very proud of you. This is a big leap for you and your future."

Oma caught Hanna's eye and smiled. Hanna smiled back, then looked away, though Oma continued to look at her.

As the sun set on that warm August evening, Hanna kissed her parents good night, just as she always had, and retired to her bedroom.

In the darkness, she closed her eyes and tried to will herself to sleep, but sleep did not come. She popped her eyes open again. Turning on the night table lamp, she slid off the bed and pulled her little brown suitcase out from beneath, then laid it on the bed. She opened it to check to make sure, one last time, that she had everything.

Folded neatly inside was a pale blue summer cotton shift, a pair of socks, and a thick woolen sweater, hand-knitted by Oma, which she figured she might need when autumn nights got chilly. Atop her clothing lay a photograph of the family, an old ten-mark bill, and a packet of Lucky Strike cigarettes that Kallehn had given her, in case she ever found a need to "bribe someone."

Suddenly, without warning, the door slowly opened, startling Hanna. Oma quietly entered. She stopped short and stood motionless when she saw the suitcase on the bed instead of a simple satchel, which is all that would be needed for a daylong excursion. Neither moved as they both stood there, Hanna looking at Oma, who stared at the suitcase. Then, as quickly as she had entered, Oma turned and walked out of the room, closing the door behind her.

The next morning, just before sunrise, with cool easterly winds bidding farewell to what had been a long and painful year, Hanna picked up her suitcase and left the house before she thought anyone had awakened. Oma, however, had risen and stood in the upstairs window, watching her daughter walk away with a determined gait, her long, dark braids falling down her back, looking more like a schoolgirl than a woman. She wondered if she would ever see her again.

$\mathcal{R}$ounding the corner by the church, Hanna walked toward the train station. Once there, so as not to arouse suspicion in case her parents would check up on her, she bought a round-trip ticket to Magdeburg. She looked behind her, half-expecting to see her parents, but was relieved to see only other passengers. She boarded the commuter car and found her seat, and the train rumbled out of the station.

She settled in and looked out the window at the passing landscape, the grassy hills and patchwork of parched late-summer sage and wheat-colored farm fields. With no map and no plan for how to make it to the border, which paralleled the tracks some twenty-five miles away, she had to focus on the best place to jump off and run westward. The train picked up speed. Rocking gently, it made its way down the track, its wheels rhythmically clicking beneath her like the second hand ticking on a loud clock.

As the train made its way along the track, her mind wandered. She was terrified to fall into the same rut Opa had been in. She had watched how during Nazi times he was constantly on edge, trying to please the authorities just to keep his job so he could feed his family. Now the same cycle was repeating itself. Why then, she wondered, had he been so eager to push her toward that same fate? It was simply not her destiny, she had decided. She was young and deserving of a future. Hadn't Opa always inspired her to have big dreams and seek adventure? And Oma had instilled in all her children a clear sense of right and wrong, taught them to be true to themselves and live up to their potential. And so, she had made up her mind. With no way of knowing what impact her escape would have on her family and believing that somehow their separation would only be temporary, like hundreds of thousands of others who were making the same decision for similar reasons, she was choosing to escape.

Snapping from her thoughts as the train started to slow in a remote rural area seemingly in the middle of nowhere, she grabbed her bag, intending to make a move for the door, when she noticed a circle of other passengers slowly standing up and starting to collect their things. Though they carried a basket with food, they also carried various valises and satchels and Hanna's instinct told her that this group was not out for an afternoon picnic.

Hanna too stood and whispered to the woman standing closest to her, "Are you going to the West?" The woman eyed Hanna, then ignored her. When the group got off at the next stop, Hanna moved with them. Quickly and quietly they boarded a two-car train that stood on the adjacent track at the small rail junction. Used primarily for farmers to transport agricultural supplies and crops, on this day the group was making use of it to transport themselves to the border.

No one said a word. The farmers adjusted their sacks and bales to make way for the group. Knowing exactly what the group intended to do, they looked the strangers over, the hopeful escapees, trying not to make eye contact, looking down at the floor or out the window as the train jolted and began moving westward. From her seat on the wooden bench across the aisle from Hanna, the woman looked her over. After a few minutes, she got up, approached Hanna, and wedged herself in the seat next to her.

"What do you want?" she asked point-blank.

"I want to go with you," Hanna whispered.

"How much money do you have?"

Hanna tried to hand her the ten reichsmarks.

"Those are completely worthless," she scoffed. Then she came in closer, "I'm risking my life to do this. The others have paid a lot of money to get out," she said, shaking her head.

Hanna pleaded.

"Listen," the woman cut her short, sensing in Hanna a trouble-maker who might blow their cover. To keep her from causing a scene, she whispered, "When we get off at the station, wait for me on the bench. I'll come back for you."

Hanna did not believe the woman would return for her and feared being left behind so close to the border, but she nodded and thanked her. The woman retreated to her seat. Other than farmers and work-ers greeting one another as they boarded the train and bidding one another farewell as they disembarked at various points along the way, no one else said anything else for more than an hour as they passed through various fields and small farming villages. Somewhere near Völpke, the group disembarked.

The woman silently pointed to the bench. Hanna dutifully went to sit on it and watched the group disappear into the forest. As soon as the last of the group was out of view she got up and followed, keeping her distance, hiding behind trees without letting them out of her sight.

Up ahead, they walked silently and swiftly, single-file through the dense woods and tangled underbrush, halting in the woods when their guide stopped, moving when she moved. Eventually they came upon a dirt path, which they followed to the back side of a barn. Just as they turned the corner, bypassing the barn, from a distance came loud shouts barked in Russian commanding the group to stop.

Shots rang out and someone screamed in pain. Hanna pulled herself behind a tree and froze. The exploding blast of more gunfire, another scream, chaos, pandemonium, and the smell of sulfur from the gun blasts filled the air. In the midst of it, Hanna, still out of view, ran toward the far side of the barn and frantically tried to open the heavy wooden door. The door was tightly shut, so she raced to an adjacent old barn, her heart beating furiously.

Once inside, she closed the door, then turned around to find herself face-to-face with a tiny, frail, doe-eyed woman who sat on a woodpile, alarmed at the commotion and the intrusion. Seconds later, a Soviet soldier yanked open the stable door shouting something in Russian, ready to use the butt of his rifle. Hanna stumbled back.

The woman shielded herself, preparing to take the blows, pleading in broken Russian, "Have pity on a poor old woman and her niece."

He stopped short. Looking around the barn, he noticed the woman's belongings strewn over straw stacks, clothing hung over wooden beams and atop a pile of straw, a man's work jacket here, a cooking pot there, bedding, suitcases. He eyed the woman and then Hanna.

"Please," she continued. "We mean no harm. I am sick and my niece takes such fine care of me." He paused, looked over the hovel, then lowered his rifle. Apparently convinced of her story, he left.

Slumping back onto the woodpile, the woman sighed. She was a refugee who had lost her husband in the war and was living in the barn with her son. For the time being, he had been able to keep his job at a sugar factory just across the border in the West and, therefore, had a special permit to travel back and forth. As the woman explained, Hanna nervously peered through a crack in the wooden door, watching the soldiers haul away the survivors of the group.

The woman's son, a young man in his twenties, arrived home later that afternoon with a loaf of bread and a bottle of milk, which he handed to his mother. He agreed to help Hanna flee through the border point he crossed every day. He asked Hanna how much money she had. She gave him her reichsmarks and the package of Lucky Strike cigarettes. He gave her ten West marks in return and said, "Remember: you are my cousin and I am allowed to cross. So don't be nervous. I know these guys. Stay calm, act natural. Let's go." The woman wished her well, and the two set out for the crossing.

As they approached, two Soviet guards, both with self-rolled cigarettes dangling from their mouths, greeted the young man like an old friend.

"*Vanja*, my little brother!" one Russian guard exclaimed in broken German, smiling broadly and slapping him on the back. "Hey, who's your pretty girlfriend?" He told the guard that Hanna was his cousin and that she wanted to go and visit their grandmother in the West for a few days. As he talked, he produced the cigarettes, which delighted the guards, who cheered as one grabbed the pack and swiped it under his nose, comically inhaling its aroma. The two "cousins" bid the guards farewell and, with a friendly wave, slipped off into the forest.

Once on the dirt forest path, he pushed her off, telling her to walk straight down the path and eventually she would reach the West, and then he left her to return to chat with the guards. She walked off slowly at first, nervously, wanting to run, but determined not to stir up attention. Once she came into denser woods, though, fear and adrenaline took over and she took off, sprinting deeper and deeper into the forest. She only stopped running when she saw the back of a white sign up ahead. When she reached it, she walked around to see what was written on it.

With her back momentarily to the West, she looked one last time into the East and at the sign, which read: "Warning: You are now entering the Soviet Occupied Zone." Turning back toward the West, she walked into her new world.

*I*n Schwaneberg, when Hanna failed to show up on the last train home, Opa exploded. Oma grew quiet. In Seebenau, when Kallehn got the news, he grinned.

PART TWO

5

TWO CASTLES
OUT OF THE WHIRLWIND
(1948–1949)

The soul has illusions as the bird has wings.
—*Victor Hugo*

*H*anna emerged from the pine forest in the West Zone weary and shaken. With no sign of life in either direction on the deserted country road that lay before her, she chose a path and, suitcase in hand, started walking.

After a distance, she spotted a horse-drawn wagon. As it neared, she flagged it down, then asked for a ride with a farmer, who recognized her as a *Flüchtling*, a refugee. She climbed up and sat beside him and only then felt relieved. She took in a long-awaited deep breath of fresh air. He smiled and patted her on the back in gentle congratulations. Then he turned back to the road and shook the reins.

As she rode atop the farm wagon, she vowed to never look back. But then, unexpectedly, visions of Oma and the rest of her family popped into her head and suddenly she felt light-headed. She began to sweat, her heart hammering as the first prickles of panic crept up her neck. The farmer noticed none of it. Then, as if in a nightmare, she could sense her family, her childhood, and everything she had ever known being swallowed up and finally disappearing into a black hole.

She tried to shake off the suffocating feeling, a survival instinct took over and she willed herself to focus on her next move. She had to find a place where she could lie low and remain anonymous until she turned twenty-one in a few months' time. Then she would be able to register as a legal citizen of the West, and finally shake the long arm of the law. Until then, she had to hide.

Before the farmer turned off the main road, she thanked him, climbed down, and continued on her way, intermittently hitchhiking and walking for the next several hours. By nightfall, with only the sliver of light of a quarter moon to guide her, she came upon the sleepy little village of Dettum, where she slept overnight on the floor in the corner of a vacant building.

The next day, a plump, aproned village matron spotted her walking into the village. When she learned Hanna was from the East, she took her under her wing, leading her first to the *Gasthaus*, where she fed her a bowl of vegetable beef soup and asked about her plans. After learning Hanna was on her own, and needed to earn money, she took her by the arm and led her to the edge of the village to meet the Schneiders, a young farmer's family who needed a housekeeper and nanny, and were willing to look the other way at Hanna's illegal status.

*B*ack in Schwaneberg, it wasn't long before members of the newly minted East German police, the VoPo, paid Opa a visit.

"If she is not dead," the police told him, "she will contact you. And when she does, you will contact us."

Opa understood. His job and the well-being of his family were in jeopardy because his daughter had committed the worst of all crimes against the state by depriving the Soviet Zone of a healthy, able-bodied citizen it needed to help rebuild the country.

But Hanna did not contact her family, and, in fact, hoped to disappear into complete obscurity. With each passing day, Oma and Opa wondered what had happened to their daughter. Had she been shot at the border? Had she made it out? Was she dead? Was she languishing in a prison somewhere?

Spurred on by the notion that the window for escape was slowly closing, thousands made a run for the West, taking their chances on getting caught. Though it was still relatively easy to cross in some outlying areas, security on the border was increasing every day. Hanna was one of the lucky ones who made it out. Many were not as fortunate, and by now, young and old, men and women, families with children, were shot or arrested and sent off to prisons that were rapidly popping up throughout the East Zone.

Prisons now saw a steady influx of those who tried to flee, those who challenged the system, and many who were swept up arbitrarily and charged with wrongdoing. Inside the notorious Hohenschönhausen and Bautzen prisons, and even Buchenwald and Sachsenhausen, former Nazi concentration camps that had been turned into East German prisons, the Soviets and communist Germans wasted no time extracting confessions often based on fabrications, misrepresentations of the truth, or complete lies.

"Traitors" of all kinds were brought before kangaroo court Soviet tribunals, where defendants were presumed guilty and sentenced to hard labor or death. In the first years, the Soviets would execute hundreds of East Zone citizens.

*H*anna worked for the Schneiders for nearly three months, cleaning, cooking, and taking care of their young son. All was well until a farmhand who had had a tiff with Mr. Schneider tipped off the police that an illegal was working for the family. Compelled to investigate, but reviling his role in returning escapees to the East, a West Zone policeman reluctantly paid the Schneiders a visit. He wrote down Hanna's identification information.

"Please," she pleaded, "in two weeks I will be twenty-one years old and then I will be legal." The policeman held Hanna's gaze. Then he nodded, smiled at her, and snapped almost comically to attention as if he were about to make a profound declaration.

"Fraulein!" he announced with a mock, exaggerated air of authority. "You are a minor and in the West Zone illegally. By law I am obligated to take you into custody and deliver you back to your guardians in the East. I will be back in *two weeks* to pick you up." With that, he smiled, bid good day, and walked away.

Two weeks later, the East Zone police and my grandparents received an official cable announcing that their daughter had become a legal resident, and had chosen to remain in the West.

*T*he local VoPo commandant summoned Opa and Oma to their new headquarters, the building that had once housed the local Gestapo office. His disappointment palpable, he stared coldly at Opa across the desk, trying to intimidate him. He said little besides "how unfortunate that your daughter has decided to flee," then shook his head and told them to go.

It was clear the stakes were now higher. Fearing the impact that Hanna's escape and refusal to return would have on his family, Opa threw himself into his work, hoping to regain favor with the authorities. Oma began to worry about Opa's ability to cope with the burdens he now carried, the stresses of trying to paint himself as a good communist, trying to provide for his family, but now with the added label of "father of a criminal."

The rest of that December passed and, knowing they were being watched by the authorities, Opa cautioned Oma and the children to avoid drawing any attention to themselves. That year Oma thought it best to lie low and refrain from celebrating Christmas in any way whatsoever.

In the West, on Christmas Eve, Hanna became terribly homesick. She went with the Schneiders to church and, when the organist played "Stille Nacht," "Silent Night," she sobbed uncontrollably in the midst of the service, to the utter embarrassment of the Schneiders and the annoyed looks of the other worshippers. Unable to control her tears, she ran out of the church and all the way home.

After the new year, Hanna gathered up her wages and her few belongings, chopped off her braids with a pair of Mrs. Schneider's sewing shears, bid farewell to the Schneiders, and took off on a train heading farther west, bound for the castle city that was calling her. In Heidelberg, she hoped to disappear into the crowds of a bigger city.

She arrived in Heidelberg at sundown and spent the first night sleeping on the freezing floor of the main train station with little more to keep her warm other than a secondhand wool coat that Mrs. Schneider had given her upon her departure. The next morning she awoke to a policeman pushing her off. She rose and went to find the Heidelberg Castle.

As the chill of the Cold War winds blew in, the United States, United Kingdom, Canada, and other Western countries began to realize the need for a cohesive military alliance against the Soviets. Hoping to discourage acts of aggression by the Soviet Union, the North Atlantic Treaty Organization (NATO) was created with all parties agreeing to mutual defense of one another in response to an attack by any outside force.

$\mathcal{M}$eanwhile, while shoots of democracy were taking root in the West, the East German regime took control of the media, censoring all the news and information that reached the eyes and ears of the citizens of the East Zone. Amid a budding propaganda and misinformation campaign, and with strict consequences for tuning in to radio signals from the West, Oma, Opa, and the rest of the family were not aware of what was happening outside the East Zone. Few if any of the villagers had heard anything about the establishment of NATO or anything else happening in the West. Like everyone else in the East, however, they sensed that the Soviets were intent on severing ties with the West and drawing the East Zone further into isolation.

Firmly convinced that the success of communism in Germany lay in the hands of the next generation, the Soviets launched a youth movement. Chosen to lead the movement was thirty-three-year-old hard-liner Erich Honecker.

A Party stalwart since the age of eight, Honecker had joined the German communist movement at fourteen. During the Nazi years, he was sentenced to ten years' hard labor for participating in communist activities and for refusing to repudiate his ideological convictions. Freed by the Red Army in 1945, he became one of the first German Communist Party members in Soviet-occupied Germany.

Viewed as a trailblazer and a visionary, physically the small Honecker was hardly the image of Soviet might, and his humorless demeanor hardly made him likable, but what he lacked in charisma, he made up for in ideological ambition. Staunchly driven to reshape the Soviet Zone, he set forth to transform the country's youth. Some twenty years later, Honecker would be rewarded for his contributions to his country by becoming the leader of East Germany.

Honecker quickly set to work establishing the FDJ, or Free German Youth, and the Young Pioneers. Much like the Soviet Komsomol, the programs were grounded in propaganda and portrayed communist society as a rich and fulfilling, all-benefiting utopia, its youth icons of patriotism. Scouting and sports activities aimed to win over the children's hearts and minds and recast them as future revolutionary leaders of their country. Teachers like Opa were forced to vigorously promote the program and encourage the children to participate in the meetings. Parents soon realized that participation in the system came with benefits. It would lead their children to opportunity and advancement in education, whereas lack of participation, simply, would not.

In Schwaneberg, the first gathering of the FDJ and Young Pioneers was quite a spectacle. Like a preacher welcoming Sunday parish-

ioners, Mayor Boch stood outside the community house greeting the
children with smiles and hearty handshakes, making a mental note
of who did and did not show up. Hanna's siblings all showed up,
Oma and Opa having directed their children to go, knowing it was in
the best interests of the family.

Klemens was initially reluctant to take part, but Tiele was intrigued, having decided that the word "free" in an organization called
the Free German Youth might mean the group would have something worthwhile to offer. The activities would also give her a chance
to socialize more often with friends and classmates, which, in the
rigidness that the school environment had become, she longed to do.
At just thirteen, Manni was still too young for the FDJ but couldn't
wait to join the Young Pioneers. To him it all sounded like great fun,
especially after the long, arduous years of war. With her oldest son
a teacher and three of her children off to join the communist youth
movement, Oma remained at home with the littlest ones, Kai, Helga,
and Tutti.

Nestled into the Odenwald Mountains, the Heidelberg Castle over-
looks the romantic old city of Heidelberg, which lies sprawled out
in the green Neckar Valley below. In a gentle snowfall on a frigid
January day, Hanna made her way from the train station through
the downtown cobblestone city streets. The aroma of freshly baked
bread steered her into a bakery, where she purchased a *brötchen* bread
roll, then emerged back out onto the street into the bustling morning crowd, which parted for the clanging streetcar plowing through.
Moving with the masses, she stopped in her tracks when, off in the
distance, she spotted the castle. Shoppers and workers pushed past,
bumping and jostling her as she stood rooted, staring at the fortress,

mesmerized by its majestic presence, which had come to symbolize her quest for freedom.

She took in a long, deep breath. Then she crossed over the old stone bridge that spanned the Neckar and slowly made her way up the hill.

*F*rom the castle kit model that she had helped to assemble as a child, she knew much of the architecture by heart, including the design configurations of the various structures, towers, turrets, and spires, the locations of the archways and alcoves, and inside, the hallways and staircases. Her job had been to piece together the south towers with Manni, who had been particularly fascinated with the tiny drawbridge that actually opened and closed.

Now she stood before the real palace, taken aback at its massive size and construction. Her eyes panned the length of the great, red sandstone structure. Once just a child's chimerical fascination, it was far bigger than she had ever imagined.

She wandered the grounds, thinking about her family, and about the journey that had brought her to this point. A flood of emotions consumed her, her heart aching for them to be there with her.

From the hillside, she turned to look back down toward the city. Beyond the valley, endless blue sky stretched as far as the eye could see. As she scanned the skyline, a feeling of peace washed over her. One chapter in her life had closed and another was just beginning. At a little shop nearby, she bought a postcard of the castle and tucked it away.

*D*eep inside the East Zone, another castle, almost as grandiose and imposing as the Heidelberg Castle, was receiving a lot of attention for an altogether different and very sinister reason.

The Hoheneck Castle too sat high on a hilltop, but in the Erzgebirge Mountains, dominating Stollberg, a lovely, picturesque German town that lay in the lush, green valley below. Both fortresses were built in the thirteenth century and had been home to great royalty and an eclectic array of renowned European dignitaries and distinguished luminaries. But unlike the Heidelberg Castle, which was open to tourists coming from around the world to learn about Germany's rich history through its renowned Gothic and Renaissance architecture, the Hoheneck Castle was closed to visitors, and in the process of being turned into a women's prison.

Women of all ages—teenagers, young mothers, some with children, middle-aged and older women, even pregnant women—filed in. Some had no idea why they had even been arrested. Others had been charged with various infractions and crimes: attempting to flee the Soviet Zone, inciting dissent, participating in secret underground organizations that conspired to conduct subversive activities against the regime. Many were innocent of the charges, but the regime could take no chances.

Terrified women and girls shuffled through the heavy iron gates into the creepy castle, one behind the other, where they were met by brutal guards, then stripped and thrown, in groups of thirty, into pitch-black concrete cells designed to hold four. There, skin to skin, in total darkness, with no room to sit, they were made to stand in knee-deep freezing water for days on end in dank, poorly ventilated chambers until they simply passed out from exhaustion and despair. From there, they were released to overcrowded prison cells, where they awaited their summons to appear before a Soviet tribunal. During interrogation they were beaten.

Some were given a chance to redeem themselves by working for the police, spying on and extracting information from their fellow prisoners that could be used to incriminate them and others. Those who refused to spy were often promptly executed. Others were sent to hard labor or to further torture in the castle dungeon, their mountaintop screams muted to the inhabitants of picturesque Stollberg in the valley below.

6

A SISTER BORN IN THE EAST
THE STASI TAKES CONTROL
(1949–1952)

*It has to look democratic, but we have got
to have it under control.*
—*East German leader Walter Ulbricht*

*T*hroughout the East, by and large, people had no real understanding about what was happening in the prison system. In Schwaneberg, Oma and Opa carefully watched over their children. Knowing they could not risk another child challenging the regime, they set a good example of conforming, abiding by the law and following rules, and instructed their children to stay out of trouble. In an effort to preempt any ideas the children might have about fleeing, Opa gave clear warnings.

"It's extremely dangerous and anyway, your life is here with your family." Just for good measure, he added, "No one goes anywhere." With the stakes now higher, they had to make sure their children remained amenable and compliant.

Unbeknownst to Oma and Opa, however, just a few months earlier, nineteen-year-old Klemens unsuccessfully tried to follow in his sister's footsteps. The summer before starting teachers' college, he offered to go up to Seebenau to help Kallehn on his farm. One day, he made a run for the border, but he was caught and taken back to the Soviet Commandatura to be processed into the prison system. Someone ran to the farm to tell Kallehn, who was eating his noonday meal in his kitchen. He got up from the table, charged out the door, stormed to the Commandatura, and demanded to see his grandson.

A Russian soldier brought Klemens forward and Kallehn smacked him hard in the face and angrily hollered, "What are doing? You come home with me. I need you in the field!" With that he dragged Klemens out of the building, leaving the soldiers shocked and speechless, but no doubt saving Klemens from a term in the prison system.

*H*anna had no real sense of how lucky she had been to escape. Completely oblivious to the terror young women like her were facing at the Hoheneck Castle, she spent the next few days exploring Heidelberg.

Thrilled at the idea of getting a Western education, she made her way to Heidelberg University and proudly offered her new West German citizenship papers to the registrar. But there was a problem. Because she had not completed high school, she was barred from enrolling for university classes. Stunned, she pleaded, explaining that she was a refugee from the East and had been only one week shy of obtaining a high school diploma when she had fled. Surely the registrar could empathize. Rules were rules, he retorted, dismissing her with a flick of the hand and calling for the next in line.

Disappointed but undeterred, she walked to the local community college and signed up for English classes. After paying one month's tuition, she then searched for a flat. A six-by-eight-foot room with a dirty, battered old couch was all she could afford. With the money she had left over, she secured the room for a month.

With her last funds spent, she needed to find a way to make money, but without official work credentials or a high school diploma, she was limited to the lowest-paying jobs. Over the next year, she worked as a piano player in a downtown city bar, a hat-check girl, a house-keeper, a nanny, and even went door-to-door selling underwear. She often went without food in order to save money for rent and tuition.

One day, while standing before her English class making a presentation, having eaten little in the preceding three days, she fainted. A classmate jumped to take her to a doctor, but when Hanna came to, she refused, knowing that she couldn't afford medical care. All of her classmates knew the skinny little East German refugee was too proud to accept handouts and was trying to make it on her own. Soon she began to find sausage, bread and butter, and other gifts left from anonymous donors in her schoolbag.

In the little spare time she had, she learned to type on a classmate's typewriter and studied English shorthand. After a few months, she was able to make a down payment on her own typewriter. But at the beginning of each month, when rent was due, she always took the typewriter to the pawnshop to be hawked, but told the pawnbroker not to sell it because she would be back after the first of the month to get it back.

In Schwaneberg, Oma and Opa still had no word from Hanna. It had been six months since she had escaped. For Oma, the time was marked by hollow days and she became distressed not knowing what had become of her daughter. For a time, the mood at home was reserved and quiet, except for Opa's outbursts, which were becoming more frequent. There was no discussion of Hanna's flight to the West; it was a painful and off-limits topic. The children knew not to speak of their sister who had fled, yet they missed her terribly, and the youngest ones, especially, couldn't even fully comprehend why she had left, and felt as if she had simply vanished from their lives. Although Roland had chosen another path, as painful as it was to bear the separation from his beloved sister, by now he had dedicated himself to his career and building a future within the system.

*S*ome eight months after Hanna had fled, and to Oma's great relief, the family in the East finally received word of her whereabouts in the form of a postcard from Heidelberg.

Oma held it like it was a winning lottery ticket. The children gathered round. On one side was a beautiful color photograph of the Heidelberg Castle; on the other Hanna had written that she was safe, was working hard and studying, had her own flat and was eating well. She ended her note, "Papa, the Heidelberg Castle is just as wonderful as you said it would be."

Opa read the postcard, noticing there was no return address.

Seven-year-old Kai asked, "Is Hanna ever coming home?"

"She'll be back," Opa answered.

The rest of the children perked up looking at him, wide, inquiring eyes all around. "It's not that easy out there," he said, looking at each of them separately, squarely in the eye, "to be on your own, so young and all alone with no one to help you. Here at home you have everything you need, a family that loves you. Out there, who knows how she is making money to live. And there are so many dangers. She is surely having a hard time. I'm sure she regrets her decision."

Oma knew that Hanna was struggling, but she also knew she was not coming home, so, unbeknownst to Opa, Oma put in an application to travel to find Hanna in Heidelberg. Two months later, the authorities provided her an answer: DENIED.

Not long thereafter, Oma's heartache for her oldest daughter temporarily abated when she learned that, once again, she was pregnant. At the age of forty-four, she prepared to give birth to her ninth child.

Deep inside the East Zone, the Berlin Airlift was coming to a close. By April 1949, hundreds of daily Allied flights had successfully supplied West Berliners with all sorts of lifesaving provisions, including flour, coffee, milk, cheese, even coal and gasoline. After nearly a year, the airlift had delivered more cargo than had previously flowed into the city by rail. The Soviets, who had repeatedly claimed the resupply effort would never work, were humiliated. Eventually Stalin conceded defeat and lifted the blockade as ground, rail, and air access routes were reopened for regular business.

The Western powers' victory over the Soviet blockade reassured the people of West Berlin that the United States, United Kingdom, France, and eventually the whole of NATO would not abandon them and would keep their promise to defend their city against the Soviets.

*M*eanwhile, the United States and the Soviet Union worked to build their nuclear arms capabilities. The U.S. program was initially at an advantage, having already developed nuclear weapons detonated in Japan to end World War II. The Soviets jump-started their own program by stealing secrets from the United States. Then in August 1949, the Soviets surprised the Western world, which believed the

Soviets were not yet nuclear-capable, when they detonated an atomic bomb in the remote steppes of Kazakhstan.

The competition for world supremacy had begun. Changed from a weapon to be used to end a war, nuclear weapons suddenly became a tool of containment, a mechanism that both sides could use to keep the other at bay, fueling a rivalry that became a competition for superpower dominance.

The nuclear arms race, which would be at the heart of the Cold War for the next forty years, was on.

*O*n October 7, 1949, East Germany was officially established as a satellite state of the Soviet Union, with East Berlin chosen as its capital. A few months earlier, West Germany had been established, with Bonn, located far west of Berlin, named as the seat of government.

The Soviets installed East German leaders but retained de facto control. Walter Ulbricht became East Germany's first leader. Ulbricht preferred a title in line with his Soviet counterpart, Joseph Stalin, so he officially became the general secretary of the Central Committee of the Socialist Unity Party of Germany, and the leader of the Communist Party.

A secret police was formed. Trained by and taking its lead from the Soviet Union's KGB, the East Germans established the Ministry for State Security, the MfS, or Stasi for short. The Stasi was charged with preserving the security of the regime and conducting espionage in country and abroad. Its most sinister task, however, would eventually become the wholesale manipulation and control of the citizens of East Germany.

Erich Mielke, a former member of the brutal Soviet secret police, became the Stasi chief. In just a few short years, Mielke would become the most feared and hated man in the country, equipping his agents with a full range of disturbing physical and psychological torture tactics, urging his Stasi officers to execute if necessary, even without a court judgment. Ruthless and unaccountable, from the outset the Stasi operated above the law, relying on clandestine operations and using fear tactics and intimidation to reach its goals.

Though the Soviets retained ultimate decision-making authority, by early 1950, East Germans had taken over all major administrative and functionary posts in the government, as authorities and bosses at factories and schools and as guards along the border. Scores of East German citizens were recruited to become Stasi agents.

Recruited from all parts of society, including the FDJ and police force, new Stasi officers and employees were top leaders in the Party, the most promising young communists in society, as well as tradesmen and technicians. Many were pulled from underprivileged or proletarian segments of society. After being vetted for their political loyalty and intellectual capacity, they attended intensive training programs in which they learned Marxist theory as well as how to pressure, brainwash, and manipulate their targets, the people of East Germany.

The Stasi organization's goals grew quickly. Stasi agents operated broadly and without bounds. They infiltrated West German and other foreign intelligence activities. Inside East German borders, they targeted anyone they believed could be a threat. Before long, there were separate departments for surveillance, blackmail, arrests, and torture. In addition to monitoring "the class enemy," anyone who opposed the regime, they also began to keep track of those who could be a potential danger and those they could manipulate to be of use to them in the future. With limited initial manpower to keep track of everyone they wanted to monitor, they launched a campaign to have East German citizens spy on one another.

*T*hough the Stasi had not yet set up a presence in Schwaneberg, the local authorities and VoPo police continued to keep close tabs on the family. In light of his daughter's escape, the authorities and Mayor Boch began to take advantage of Opa's vulnerability, manipulating his influence with the villagers and leveraging what they called his "black marks" to force him to do exactly what they wanted for the regime.

But the Schwaneberg authorities realized that they had a problem the day Opa spoke up to represent the plight of the local farmers who, in the early 1950s, were being forced to give up their land to the state.

The order was a devastating blow to farmers, many of whom, like Kallehn, whose land had been in their families for generations, were now directed to hand it over to be developed into state-controlled agricultural collectives. Some farmers rebelled, setting in motion another law that forced those who did not cooperate to give up their livestock and then, if they persisted, everything they owned of value, including their most prized personal possessions. Those who continued to resist were arrested and imprisoned. In Seebenau, Kallehn learned to keep his head down and avoid the authorities, but it was only a matter of time before he would be confronted too and the notion of giving up his land continued to pain him deeply.

One day, the farmers in and around Schwaneberg assembled to express their outrage in a town meeting. The community leaders, including Opa and Mayor Boch, attended the gathering, as did scores of farmers. Since well before the arrival of the Soviets, it had long been Opa's role and civic duty as one of the village leaders to listen to the concerns of the people and represent their interests to the local government, which Opa did with great pride and genuine concern. If anyone could have an impact, the farmers believed, it would be Opa, who had an excellent track record for initiating action and making things happen on the community's behalf. Confident that his recent membership in the Communist Party was evidence of his support for the government, Opa expected those in charge to respect and take into account what he had to say. But when he confronted the state on behalf of the farmers over seizure of their land, and for villagers regarding their private property, the authorities simply ignored him. Opa was stunned. Then he did something that made him a marked man.

One of Walter Ulbricht's first addresses to the people as the new leader of East Germany encouraged them to believe that the Ulbricht government intended to use democratic principles to establish the communist state.

"The nature of a democracy," he said, "consists to an important degree in the right of the people to criticize problems and mistakes. I ask you," he continued, "to let the government know immediately when you see serious problems or mistakes that stand in the way of our great community endeavor."

Perhaps naïvely, Opa chose to take advantage of this so-called open-door policy, either hoping or truly believing that the Ulbricht administration sincerely sought feedback, especially from community leaders, to help them mold and shape the new society. Livid at what he saw as disrespect for the village leaders and neglect for the interests of the farmers, the backbone of the country, and taking it as a personal affront to his position, Opa decided, unbeknownst to Oma, to appeal directly to the leader of East Germany, Walter Ulbricht himself, by writing him a personal letter in which he presented a compromise between the government's plan to force agricultural collectives and the farmers' interest in keeping their land. After all, had not Ulbricht himself encouraged people to speak up where they found problems? Several days later, the authorities came around.

They warned Oma and Opa that they were walking on thin ice. Oma, shocked at what Opa had done, took the admonishment stoically, but Opa could not. He responded by pouncing back, raising his voice, and objecting to their hypocrisy. The authorities were astounded at his audacity and called him a troublemaker.

After that, Opa was marginalized by some of the villagers, who, even if they believed he had been right to speak up, also knew that continued association with him risked getting them into trouble. As villagers observed Opa's predicament, many felt conflicted between their long-standing respect for him and his apparent need to stand up to the authorities, but for certain, almost everyone began to see him as an example of how not to behave. Some who had always held their school headmaster and teacher in high esteem began to wonder why he insisted on confronting the system, and they believed his days as a village leader were numbered.

At home, Oma pleaded with Opa to control his temper and to keep his outbursts and criticisms of the government to himself.

*I*n 1950, the family in the East finally received a letter from Hanna that showed a return address. In Heidelberg, Hanna was overjoyed to receive her first news from home. Oma's letter was brief and lacked detail; clearly she had kept her true thoughts to herself so as not to draw attention from the authorities who were likely reading the family's outgoing mail. There was one bit of big news that came as a complete surprise. Oma had given birth to another child.

One year after Hanna had escaped, in July 1949, the same year that East Germany had officially been established as a new and separate state, Oma wrote that Hanna had a new little sister.

From the start, the tiny newborn with a shock of black hair and wild, flashing eyes had a remarkably hearty cry. She was a lively, alert baby. Her brothers and sisters doted on her. Oma was refreshed with a new sense of purpose. She named the baby Heidelore and called her Heidi.

*E*ast Germany's new leaders tried to gain wider support by appealing to the German psyche and trying to win people over. Now villagers were allowed to keep the proceeds of their small backyard gardens, and one could celebrate Christmas provided it had no connection to religion. Those insisting on keeping their ties to the church were allowed to do so, though later they would be marginalized for their association. And more emphasis was given to engaging the youth.

Membership in the FDJ and Young Pioneers was officially voluntary, but in actuality it was required when one considered the consequences for not participating, the regime having made it clear from the beginning that there was no real choice for those who cared about their future. Worked into the framework of the school day, it was hard to avoid membership in the youth program. Not joining would draw unwanted attention since school officials were instructed to report refusals and disinterested youth to the authorities.

Though designed to be fun, the youth programs were actually engineered to embed orthodox communist doctrine. Socialist teachings with a revolutionary bent infused both blatant and subliminal propaganda messages meant to graft a new mind-set. Engaging activities and seemingly nurturing lessons on becoming happily fulfilled members of society, would in time, the regime hoped, build an entire generation that would see the East German communist regime reach its goals. At schools and youth meetings, children learned to celebrate communism and to report others for their antigovernment thoughts, comments, and jokes that were out of line with regime thinking.

Children were encouraged to report rule breaking at home, such as whether their parents listened to forbidden West German radio or made disparaging remarks about the system. Vigilance in reporting others for their failings came with rewards: public accolades, special treatment, promotions in their youth group, the authorities all the while carefully noting who was and was not fully investing. In time, many children and adults would come to view these invasive behaviors as a normal, required aspect of life, of doing one's part to contribute to the country's communist development.

*I*n Schwaneberg, the youth program blossomed like a garden bringing forth new buds. Using the motto "Freedom and Friendship," Mayor Boch pulled the children in. Manni moved up to the FDJ, and at seven years old, Kai was summoned to his first Young Pioneer meeting. Like all children throughout the East, he was instructed to learn the new East German anthem and other songs that praised socialism in the East.

He came home stumbling over the words, trying to recite the Pioneer pledge: "*We love our socialist fatherland. We are friends of the Soviet Union and . . . oppose the lies of the imperial . . .* I had fun today," he broke off. "We had a race and I beat Markus." Handing Oma a paper with the creed, Kai said, "We have to learn this and you have to get me a scarf." Then he darted off to play.

It pained Oma to watch her children enter the youth movement. Nothing good, she thought, could come from the East German regime manipulating the minds of the country's vulnerable youth. She could see how such pledges filled with propaganda had taken the place of prayers and hymns in the way that they invited worship of an ignoble and sinister power. Above all, it bothered her to know that children, in particular *her* children, were promised rewards for turning against their teachers, neighbors, and, even worse, their own siblings and parents.

What will become of a country, Oma wondered, when a mother cannot even trust her own children, and they, in turn, cannot trust their own families?

Despite the fact that Opa had already tried to preempt the issue, telling his children, "Informing on your parents and one another simply will not happen in this family," there was a real concern in every family that such an act of betrayal could occur. Sadly, in some cases it did; parents saw prison terms after being turned in by their own offspring whom the regime then publicly praised and promoted for their allegiance and commitment to the cause.

At home, the children had only to look at their mother to understand that loyalty to family meant everything to her.

"We are a family and that is that," she told them. "No matter what anyone else ever tells you to do, you know how to do what is right. Do not do anything you know is wrong because someone frightens you into it. The right way is in your heart and in your soul. And that is what is most important."

In the first five years of East Germany's existence, more than a million Germans fled toward West Germany. Besides a huge loss of farmers and workers, the state suffered a significant "brain drain" of skilled and well-educated specialists, leaving, for example, towns without doctors, research institutes without scientists, universities without professors. Mass defections came from all communities and even included members of the Communist Party, border guards, and the FDJ. The chief of Soviet security himself, Lavrenti Beria, remarked that the increase in the number of defections to the West was due in part to increased hostile propaganda directed at East Germans by West German subversives. It was also attributed, he said, to peasants avoiding committing to agricultural collectives, and to the youth avoiding service in the East German armed forces. Owing to the shortcomings of the regime itself, he acknowledged a problem with the supply of food and consumer goods to the population.

Regardless of the reasons for it, one thing was clear: East Germany was hemorrhaging its much-needed intellectual and labor forces.

In 1952, the regime finally responded to its uncontrollable exodus problem by building a barrier to halt it. In what would soon become the world's most heavily fortified border, all along the East–West German divide, now referred to as the Inner German border, *Aktion Ungeziefer* (Operation Vermin) forced tens of thousands to move inland, then demolished their homes and cleared the area of trees and brush to make way for the construction of what would eventually become a thirteen-foot-high concrete and barbed-wire fence fortification along the entire 880-mile border separating East and West Germany, from the Baltic Sea in the north to Czechoslovakia in the south.

Once only protected by rolls of barbed wire and roving patrols, the border was upgraded by the installation of a chain-link fence topped with barbed wire, guard dogs, and wooden watchtowers, which allowed for better observation of a thirty-foot-wide strip of carefully groomed sand to detect footsteps along the border.

Suddenly farmers near the border were permitted to work only during daylight hours, and then only under the watch of armed guards. Though obstacles prohibiting escape faced inward, designed to prevent East Germans from escaping, the government announced to its people that the fortifications were necessary to stop the flow of "imperialist enemies and spies" sneaking into the East. Where once guards were given the discretion to fire warning shots, now they were instructed to use deadly force to prevent escapes.

Those still hoping to escape would have better luck fleeing into West Berlin, where the city was not as firmly controlled owing to postwar treaty agreements to keep open access throughout the city. And so they did, fleeing into West Berlin on foot, by car, by underground rail, through the sewers, anyway they could.

The decisive new Inner German border between East and West Germany marked in no uncertain terms Europe's division into two rival political camps, and became the foremost frontier where capitalist liberal democracy stood face-to-face with its communist nemesis. Winston Churchill's reference to an Iron Curtain descending across the continent had symbolized an ideological separation of East from West, but now an actual physical barrier scarred the landscape, perversely separating the two sides in a concrete manifestation that would become the defining symbol of the Cold War.

Oma tried to ignore all the talk of border lockdown. She was consumed with the welfare of her children and baby Heidi. From time to time, before she went to sleep at night, she allowed herself to think about her oldest daughter, but by morning she always put aside her melancholy to face the new day with renewed vigor for the rest of the family, especially the baby. She refused to give up hope of ever seeing Hanna again, and so, in late 1952, after my grandparents had not drawn any negative attention to themselves in over a year, she applied again to travel to the West.

In the West, Hanna's English had improved and she was hired as a bilingual secretary at the U.S. Army headquarters in Heidelberg. She wrote a letter in which she told her family that she had landed a dream job with the Americans.

Now easily able to pay her bills, she found a flat in a better neighborhood across the Neckar River and had even begun to travel a bit. In a letter to her parents from London, she wrote about seeing the River Thames and Big Ben, beefeaters and Windsor Castle, recalling photographs from Opa's picture books.

In the East, Oma received that letter, but noticed that it had been opened before it had reached the family. Nevertheless, she was at peace knowing that Hanna was standing squarely on her own two feet and finding her way.

Police authorities continued to keep a close watch on the family, in part by scouring their inbound and outbound letters, opening, reading, and not bothering to close them, or simply not allowing post mail to go through to the family. Opa poured himself into his work, trying to make up for his disappointments to the state, but he now had at least two major strikes against him. Besides his guilt by association with an escapee and enemy of the state, he knew he had angered the regime with his ardent support for the farmers.

(*Clockwise from left:*) Kai, Helga, Tutti, Heidi

*A*s the family faced an uncertain future in the East, the house was alight with a bright and shining spirit.

Heidi was an unusually happy child. She had an infectious laugh, was incredibly inquisitive and spirited, and the family lavished attention upon the last-born, the little darling of the family.

As she grew, Heidi learned about her oldest sister from her siblings. By the time she was four, she could recognize the sister she had never met from photographs and took to calling her Hanna-she-went-to-the-West, which is how her brothers and sisters identified their sister from photographs when Heidi asked, "Who is that?"

Occasionally Oma would watch Heidi, contemplating her purity and innocence as she lingered over, or ran her finger along the length of, her oldest sister's image.

7

"WE WANT TO BE FREE"
A WORKERS' UPRISING
(1953)

When injustice becomes law, resistance becomes duty.
—*President Thomas Jefferson*

In March 1953, Stalin died in the Soviet Union. By June, there was rebellion in East Germany.

Propaganda continued to seep into every aspect of society, extolling the virtues of communist life. Though conditions remained dire in the East, socialist realism portrayed everyday life in East Germany as a worker's dream in a country that was flourishing and prospering. Posters and statues romanticized communism, featuring sculpted tool-wielding peasants and robust factory workers with taut muscles and determined, eagle-browed gazes, radiating confidence as they proudly faced the sun, stirring their countrymen to join the revolution.

Bold schoolchildren bedecked in scarves and uniforms with badges, saluting or carrying flowers, smiled down from colorful posters, inviting their peers to come along and join their ranks. Print media and state-produced films portrayed East Germans as profoundly dedicated and highly motivated to contribute to communist society and serve the regime. By the looks of the propaganda materials, East Germany was a workers' paradise. But that picture was far from accurate, and, in fact, things were about to get a lot worse.

By 1953, more Stasi agents were being dispatched to towns and cities throughout East Germany to spread their net wider and increase their monitoring of the population. At the same time, the regime directed East German citizens to "agitate" in their communities. Everywhere people turned, proclamations called for them to fully commit to the struggle for communism and incite others to follow, to demand more of themselves and their fellow workers than they ever had before. On factory walls, in plants, schools, and hospitals, in leaflets, newspapers, and on the radio, the leadership called on every citizen to become a poster-perfect model worker for the state and to finger those who fell short of the ideal.

Pamphlets distributed by the state gave the public specific examples to follow:

Comrade Paul Wilk is an agitator at Thälmann Plant I in Suhl. His previous work has shown that he understands how to present the policies of the party and government in a simple, forceful, and consistent way. He knows how to inspire the masses, and persuade them to join us in realizing our great goal of building the foundations of socialism. His colleagues and fellow workers see him as a good worker. They respect him.

Because of his abilities, Comrade Wilk was elected agitator by the party membership. He constantly and diligently fulfills this party assignment. Every day, he reads the party newspapers "Das Freie Wort" and "Neues Deutschland," and studies the (translated) Notebook for Agitators as well as agitation literature.

As leaders rallied the masses to become activists in their communities, the state also imposed harsher travel rules, this time within East Germany itself. Now villages that lay near the border were cut off to everyone except the residents of those areas; suddenly Oma could no longer even travel to see her parents in Seebenau.

Alarmed by the numbers of people fleeing East Germany, Moscow ordered East German leader Ulbricht to make necessary changes to get things under control, to ease up on hardships and to make life easier on his people so that his citizens would stop trying to get out. But Ulbricht simply dismissed that directive, and instead his demands on the population surged. With the East–West Inner German border sealed, increasing Stasi scrutiny, pressure to spy on one another, and greater restrictions on travel out of, and now even inside, their country, the morale of the population spiraled downward.

Tone-deaf to his citizens' concerns, Ulbricht added to their misery by launching a plan to restore the country's heavy industry. Since East Germany was losing hundreds of thousands of workers to the West, the leadership now put a greater burden on the workers who remained behind. Ulbricht demanded workers give everything they had, and promised that their efforts to meet production quotas would result in substantial improvements for the country. Should workers meet the required deadlines, by 1954, he said, rationing would be abolished and all foodstuffs and consumer goods would become plentiful and affordable. By 1955, he assured them, East Germany would be well fed, and have more meat, sugar, milk, and other basic necessities than West Germany.

Before long, the cogs of heavy machinery were in motion, as the country focused maximum attention on steelmaking, machine tool building, and mining. From the start, workers were pressured to produce far beyond their capabilities, given their limited workforce and lack of resources, technology, and the state's uncompromising deadlines; they were also being forced to produce far more for the same meager pay.

As the regime continued to increase its demands, workers grew increasingly frustrated at Ulbricht's unrealistic expectations. It soon became clear that the requirement to produce under such extreme conditions and on schedule was simply an impossible one.

Rumblings of dissatisfaction spread. Workers throughout the country stewed in anger at the increasing requirement to produce more and more as food and basic goods dwindled and freedoms were stripped away. Tensions soared. The workers of East Germany were exhausted and angry.

By springtime, realizing that their demands on workers were overwhelming them, the leadership promised a shift toward lighter industry, more trade, and a greater availability of goods to consumers throughout the East: food, clothing, household appliances that would make their lives better. But nothing came of it and instead the regime kept pressure on workers to increase their industrial output while continuing to lower their wages and dismissing their basic needs.

*B*y mid-June, workers throughout East Germany had had enough. A small group organized to represent all workers bravely confronted the state demanding better working and living conditions and free elections. Their demands were met with deafening silence.

Then on June 16, everything came to a screeching halt as workers in East Berlin simply put down their tools, came down from scaffoldings, emerged from factories and work sites, and walked off the job.

Word spread and by dawn the next day, June 17, some forty thousand construction and steel workers refused to go back to work, and staged a protest march in East Berlin. Strikes, walkouts, and demonstrations broke out in most major industrial cities, hundreds of thousands taking to the streets in towns all across East Germany. In just a short time, thousands swelled to nearly a million.

Workers everywhere flooded the streets, rallying outside state offices, demanding reforms, calling for de-Stalinization and an end to the Ulbricht regime.

"Down with the government!" the crowds shouted. "Death to communism!" "We don't want to be slaves anymore. We want to be free!"

Determined to seize control, protesters overwhelmed police, grabbing their megaphones and verbally lashing out at the Communist Party and the secret police. They tore Party notices and banners from the walls, attacked government buildings, and stormed prisons, setting political prisoners free. Though isolated from the Western world, the demonstrators hoped somehow to compel the West to come to their assistance. In the United States, President Eisenhower chose not to take action to assist the East German workers out of fear of unleashing war with the Soviet Union.

The Workers' Uprising of 1953 was violently suppressed by Soviet Forces.

Resistance spread. In Dresden, demonstrators took over a state radio station and used the airwaves to attack the country's leaders, calling them liars. In Halle, rioters took over the local newspaper offices, and in Bitterfeld, a strike committee sent a telegram to the government in East Berlin demanding the "formation of a provisional government composed of revolutionary workers."

Red Army tanks moved in, and tens of thousands of Soviet troops, East German VoPo, and secret police appeared on the streets, descending on the demonstrators in a violent crackdown to silence the masses.

When it was over, hundreds lay dead in the streets. Thousands more had been injured. Some ten thousand were arrested or detained, many sentenced to long terms in penal camps. Nearly one hundred were executed for their role in the uprising along with some twenty Soviet soldiers executed for refusing to shoot unarmed civilians.

On June 18, the East German leadership went on the air to make an announcement to its people, claiming that the rebellion had been instigated by the West. In a twisted farce, Ulbricht hailed East German workers as heroes who had saved the day, claiming they had valiantly fought off the "imperialist-inspired demonstrators" who were determined to destroy East Germany. Behind the scenes, the regime ratcheted up the authority of the Stasi secret police to do whatever was necessary to ensure such an uprising would never again take place in East Germany.

The little village of Schwaneberg remained quiet throughout the uprising. Oma listened to Ulbricht's radio announcement and turned to look at Opa, who simply shook his head.

In the West, Hanna received a letter from her sister Tiele, now a twenty-year-old kindergarten teacher. In the letter, Tiele described watching the demonstrations from her window in her second-floor apartment in Naumburg an der Saale.

"There were people everywhere," she wrote. "It was a huge demonstration. Only after it was over did we learn that the West tried to stage some kind of attack." That letter the authorities allowed to go out to the West.

The East German Workers' Uprising of 1953, the rebellion of the working class, had been decisively and ruthlessly crushed. As long as the regime remained backed by the Soviet Union, there would not be another mass attempt to rebel by the citizens of East Germany.

———

A month later the local police summoned Oma. Though her previous requests to visit Hanna in the West had been rejected, they called her in to discuss her latest application. By now it had been nearly six years since mother and daughter had seen each other, their correspondence scant, having been carefully filtered by the authorities.

Down at the station, the police official sat behind a desk. He would, he said, approve a short visit under two conditions. Warning more than asking, he said, "You wouldn't leave your family in the East and try to stay in the West, would you?" Oma shook her head no. It was true. She would never abandon her family, and besides, by now it was clear to everyone that flight would mean consequences for those who remained behind.

The second condition, he said, was to bring Hanna back or, barring that, convince her to "do a few special things for her country" in her work with the Americans. Now Oma had a predicament, but she remained calm and unmoved. Refusal or failure to comply, if she accepted the terms, would also mean further problems for the family. She weighed her options. Her desire to see her daughter won out.

Thinking this might be the only chance the sisters would have to meet, Oma looked up and resolutely added, "I want to take my daughter Heidi."

"Well then," he said, and sat back, "we have an understanding."

Oma nodded. He stamped some papers and she got up to leave. Convinced that a wife and mother of so many children was not a flight risk, she was granted a travel permit to spend two days with her daughter in Heidelberg.

8

THE VISIT
SISTERS MEET
(1954)

Both within the family and without, our sisters
hold up our mirrors: our images of who we are
and of who we can dare to be.
—*Elizabeth Fishel*

*I*n the days and months following the Workers' Uprising of 1953, the regime regained control by twisting the facts of the incident as much as they could, telling East Germans that, thanks to their efforts, the republic remained intact after the unprovoked attack by West Germany. Furthermore, they claimed that the American government had approved millions of dollars for anticommunist underground activity, and that West German subversive organizations had received a great deal of that money to finance agents in East Germany to incite dissent, with fascists and provocateurs acting on behalf of foreign and West German monopolists. West Germany reacted to the brutal crackdown and Ulbricht's statements by continuing to refuse to recognize East Germany, officially or otherwise.

The Western world, for its part, came to view East Germany with increasing suspicion as stories leaked out from behind the Iron Curtain of a regime that used physical force and psychological coercion to keep its people in line, where words against the regime were enough to have one interrogated and imprisoned.

Escapees, émigrés, and the exiled brought with them alarming tales of oppression, stories of brutal interrogations, harsh prison sentences, and no opportunity for self-defense when charged with a crime. They relayed tales of torture and psychological manipulation that fueled fear and paranoia, of grave punishments for trying to reach out to the West, and of deaths at the border. Publicly East German leaders denied the accusations of abuse, but privately they realized that their reputation was beginning to take a serious hit.

Oma came home and told Heidi the wonderful news: she was going on a trip to the West to meet her big sister Hanna. Opa asked what the terms were. Oma denied there were any, hiding the truth from him.

The night before the trip to the West, five-year-old Heidi was excited and could not seem to calm down. Nor could Oma, sleeping just a few rooms away. The next morning, anticipation had them both up with the sunrise and making final preparations.

Helga helped Heidi put on her dress, her sweater, and leggings knitted by Oma. Tutti tied bows to the ends of Heidi's braided pigtails, both of the older girls silently wishing they too could go.

On a cool October morning, under an overcast autumn sky, the family went to the train station to see them off. After settling into their compartment, Heidi looked out the window, smiling and waving at her siblings on the platform. Excited for the two, the children waved back vigorously, Manni blowing kisses as the train pulled away, but Opa had already noticed the secret police standing in the background and told the children to quiet down.

Little Heidi fidgeted and bounced around the train car, chirping Hanna's name over and over again as she slid from side to side on the wooden bench, looking out of the window at the forests and farm fields as the train moved west. Several hours later, they reached the East–West border.

Oma presented her tickets and paperwork to the East German ticket examiner. Heidi beamed at him and said, "I'm going to visit my sister!" He scowled; she shrugged, then looked back at Oma, narrowing her eyes, made a comical, mocking sourpuss face, and then broke into a big smile. As the train moved onward, Heidi stood at the window, watching as they crossed into the West. She remained wide awake for most of the day, but by nightfall, when the train finally pulled into the Heidelberg station, she was fast asleep on Oma's lap.

Oma woke Heidi and they gathered up their belongings. As they disembarked, Heidi scanned the faces of passengers rushing past, looking for the face she only knew from photographs. Oma, her heart beating fast in nervous anticipation, also searched the sea of people for her daughter. Then, suddenly, from afar Oma and Hanna spotted each other and locked gazes. Hanna ran toward them, waving her hand in the air and making her way through the crowd.

To Heidi, it looked like her sister was approaching in slow motion. Like a scene in a dream, Hanna looked like a graceful angel floating toward her. It was a vision Heidi would carry with her for the rest of her life. Mother and daughter fell into each other's arms.

Then it was Heidi's turn. Hanna, her face wet with tears, bent down and took Heidi into her arms.

"So this is my little sister. Let me look at you," she said, splaying out her arms and looking Heidi over from top to bottom.

Shy at first, Heidi just looked at her sister, then asked, "Why is everyone crying? Aren't we happy?"

The women laughed through their tears. Hanna picked up the suitcase and took Heidi's hand. "Welcome to the famous city of Heidelberg," she said. Then to Heidi, "Come along now. Do you want to see my place?"

At her flat, Hanna served Oma a cup of tea and the two reengaged, trying their best to keep the conversation light and deeper emotions at bay. Heidi set to work exploring every nook and cranny and getting into Hanna's personal things. After an hour or so, the landlady knocked on the door and angrily asked who had been flushing the toilet every two minutes. Hanna found Heidi in the bathroom, mesmerized with the flushing mechanism, something she had never before seen, only ever having known the simple concept of an outhouse.

Over the next two days, Heidi barely left Hanna's side, insisting on sleeping in the same bed with her. She held Hanna's hand every chance she could, played with her hair, carried her handbag, and constantly stared at her when they sat on the streetcar, walked along the river, or strolled in the park. Heidi became completely enamored of Hanna, smitten with what she saw as a beautiful, blissfully happy young woman with a confident smile and a warm, open, carefree disposition. Heidi quickly came to think of her big sister as an example of the young woman that she herself could one day become.

They toured the Heidelberg Castle. On the grounds of the ruins, they watched an outdoor theater production of *A Midsummer Night's Dream*. Heidi became lost in scenes of fairies dancing and flitting about in verdant woodlands. At night they watched dazzling fireworks light up the night sky. The second day, they took a riverboat cruise and had coffee at the Red Ox, where just a few years earlier Hanna had played piano for tips as she struggled to make ends meet.

On their last evening together, they took a long stroll on the Philosophers' Walk. After a while, they sat on a bench to have a rest along the gleaming riverbank. While little boats with lighted lanterns sailed on the Neckar River below, there was what seemed like miles of deafening silence between mother and daughter.

At the train station the next day, they embraced. Hanna took a long time to let Oma go. Heidi nuzzled her face in Hanna's torso. Hanna hugged her back firmly, then sank down to take Heidi in her arms and look her in the eye. Stroking her long braids, she smiled at the sweet, pure face staring back at her. Heidi did not smile back, and looked away.

Then, before Heidi could see the tears welling in Hanna's eyes, Hanna hugged her again then whispered in her ear, "Be good. Take care of our mother."

Oma and Heidi slowly boarded the train. Like birds that had been set free for a short, beautiful time, it was as if they were being put back into the confinement of their cold metal cage. They shuffled to their cabin, then appeared at the window. Hanna was heartsick to see them on the train, but masked her emotions. Oma stood behind Heidi as the two stared out the window at Hanna. Hanna stared back.

Oma and Heidi in Heidelberg

When the train began to move slowly eastward, Oma waved. Heidi, looking forlorn, put her palm against the window in one last effort to connect with Hanna. Hanna waved until she could no longer see them, and stood in place until the train was out of sight.

And so it was in the autumn of 1954, the two sisters met, one twenty-six, the other only five. It would be the only time they would ever meet during the forty-year existence of East Germany.

*W*hen Heidi returned home, the children gathered round to hear her talk about Hanna, and about what they had seen and done in the West. Everyone wanted to know about the Heidelberg Castle. Opa listened to it all, but didn't ask a single question.

The Schwaneberg police did not even call Oma in. Perhaps they already knew that she had not been successful in her task of convincing Hanna to come back or become a spy. Perhaps they figured she

Heidi carrying Hanna's handbag, Heidelberg

hadn't even tried. But one thing was for certain: they chalked it up as one more failure of the family to comply. Because Oma did not hold up her end of the bargain with the authorities, they made it clear that subsequent applications to travel to the West would be rejected.

*H*eidi asked Oma for a photograph of Hanna and put it on her nightstand. Over the next months, she emulated her big sister's gestures, recalling her manner of speech, her elegant gait, her sashay. She thought constantly about Hanna, at times even pretending to be her, walking around as if she were a stylish lady with a fancy purse, a demure glance here, then there, going to a castle, riding a streetcar, having a cup of coffee at an outdoor café. At school she did not mention her visit at all. She already knew enough not to let on that she was captivated with her sister who had fled and was living a good life in the West.

At night she often fell asleep clutching Mariechen, her East German baby doll, who, with her large eyes, sweet face, and long black braids, resembled Heidi herself. Stroking her doll's hair as she fell asleep, Heidi would whisper, "Mariechen, don't you wonder what Hanna is doing right this very minute on the other side of the stars?" Mariechen gazed back, the moonlight glinting off her blue marble eyes with their faraway stare.

9

LIFE NORMALIZES IN A POLICE STATE
A COURTSHIP
(1955–1957)

You are like stale beer!
There is no need for you any longer.
—*East German propaganda pamphlet*

*I*n 1955, at around the same time West Germany became a free state, the Soviet Union allowed East Germany to declare its sovereignty.

In the first decade of its existence, the world had viewed East Germany through the prism of Soviet domination, but now Moscow gave East Germany's leaders primary responsibility to take charge of their own affairs.

The Ulbricht regime wasted no time. Huge resources went into expanding the secret police. East German authorities took complete control of information through the censorship of virtually every form of written and spoken word, from textbooks to newscasts. Ulbricht and his Soviet-trained inner circle of dogmatic communists were

bent on proving to Moscow that they were every bit as hard-line as
their Kremlin counterparts.

_T_hat same year, in 1955, West Germany joined NATO, and the
Soviet Union followed suit by forming the Warsaw Pact. While
U.S.-led NATO was governed by a system of consensus in which
each member nation, even tiny Iceland, received an equal say, Warsaw
Pact countries strictly took their marching orders from Moscow.

Later that year, West Germany established a military, and a few
months later East Germany formed the NVA, the National People's
Army, which fell in line with the other armies of communist Eastern
Europe, under the ever-watchful scrutiny of Moscow.

At the U.S. Army headquarters in Heidelberg, Hanna was sitting
at her desk the first time the new lieutenant with the sparkling eyes
passed by.

He smiled and said hello, but she responded coolly. She was used
to American soldiers trying to catch her eye, but she had no intention
of getting involved with any GI. She had just returned from a vacation in Capri, where she had fallen in love with Italy, and had already
made up her mind to follow her dreams to one day move there. In her
spare time, Hanna even began studying Italian.

But every day, the dashing young lieutenant, an intelligence officer,
politely greeted her. After a while, she realized, something about him
was drawing her in. Then one morning he passed by and stopped at
her desk. Pointing to the large satchel she carried in place of a handbag, he asked in perfect colloquial German, "What do you carry in
that big bag you're always lugging around?"

Hanna was taken aback. She looked at his name tag.

"Lieutenant Willner," she asked, "how is that you speak such good German?"

He smiled and replied, again fluently, "Oh, thank you for the compliment. We have very good language schools in America."

Beyond his lively eyes and curiously flawless command of the language, she thought, there was something clearly special about Lieutenant Willner.

$\mathcal{I}$n East Germany, there was no letup in the number of those risking escape, and by the mid-1950s, millions had fled. Despite the sealing of the East–West Inner German border in 1952, many still remained determined to get out any way they could and turned to increasingly risky methods. Some even tried to make a break for it by swimming the forty miles in the frigid waters of the Baltic Sea, hoping to reach Denmark. Most did not make it, and scores of bodies washed up on Danish shores. Others tried to flee through other Soviet satellite countries, where borders were rumored to be less stringently controlled.

The route that offered the best chance of escape was from East to West Berlin. Still interconnected by its roads, subway, and sewer systems and still unable to be completely controlled by the East German authorities, Berlin continued to be a sieve for escapes. But there was a growing concern that one day the authorities would somehow find a way to seal off the city. Thus in the mid-1950s there was an upsurge in escapes into West Berlin.

$\mathcal{A}$s people adapted to a police state with a secret police at the helm, a heightened sense of gloom took hold in East Germany. There was an uneasy arbitrariness to it all. Throughout the country, as the Stasi beefed up its network of informants used to report on the activities of coworkers, neighbors, and even family members, it was impossible to know who might be reporting to the secret police or what stray thought uttered to an associate, friend, or even loved one might make its way back to the authorities.

Now when Oma talked to the neighbors or engaged with various people in the community, her radar was up and she wondered, as everyone did, if those she once considered friends had become informants. Once when Oma went to claim her dairy rations, the worker, whom she had known for years, reached out to give her several extra eggs. She declined them, thinking him a possible informant and the "gift" some kind of test or ruse. At his weekly Saturday-night card games, Opa, assuming that some in the group were likely reporting on whatever he said, kept his thoughts to himself and showed no hint of his displeasure with the regime.

By now, news sources were simply organs of official propaganda used to bend the truth. There was virtually no mention of the outside world, especially the West, unless it was negative.

Opa longed for news about West Germany, feeling it was his only means of keeping any kind of connection with Hanna. Listening to Western news broadcasts was punishable, considered an act of betrayal against the state; but despite the risks, occasionally Opa still tried to secretly tune in to BBC radio newscasts, which had been his favorite source of news until the regime started jamming the signal. Realizing the damage listening to Western broadcasts could have on its ability to keep control of the population, the authorities scoured the country looking to see which way home and workplace antennas were directed.

*H*eidi was a confident child, full of esprit and self-esteem. As most children are, she was also naturally curious, which the regime did not consider an asset in East German children. With the authorities trying to institute tighter controls over freedom of thought, Heidi's teacher pulled Opa aside two weeks into the start of Heidi's second grade and told him that his precocious daughter was asking too many questions.

*T*he next time Hanna saw Lieutenant Willner was early one afternoon when they were both leaving the U.S. Army headquarters building in Heidelberg.

"Where are you off to?" he asked.

"To the courthouse," she said, "to watch the trial of a Nazi concentration camp guard. I cannot believe what they say the Germans did to the Jews during the war. I want to see and hear for myself . . . and if it's true, I want to see justice done."

"What a coincidence," he said. "I happen to be going there myself. Shall we go together?"

At the Heidelberg courthouse, the American military policeman standing guard outside the courtroom asked for Hanna's identification.

"Sergeant, she's with me," Lieutenant Willner said, but the guard responded, "No Germans allowed, sir."

Lieutenant Willner was ushered into the military tribunal, leaving Hanna to find her way back to the office.

The next morning when he passed by her desk, she asked about the verdict and he said that the guard had been found guilty of war crimes.

Several days went by. By now Hanna looked forward to the lieutenant's daily greetings. But then he was nowhere to be seen. After a
week went by with no sign of him, a postcard arrived from Berlin. He
was there on a business trip and wrote that he was thinking of her.
He signed the card, "Eddie."

When he returned, she noticed they seemed to have more chance
encounters. Then he invited her to the movies at the base theater and
she accepted. That date was followed by an invitation to dinner, and
after that, the two began to spend more and more time together. He
asked her about her life. She told him that she had fled East Germany and about the family she had left behind. He told her little
about himself, mostly talking about his life in the army and sharing
photographs from his travels to exotic places like India and Japan.

Over time Hanna pressed, wanting to know more about his family,
where he was from, where in America he had gone to school. Finally
and reluctantly, Eddie opened up. He was a German Jew and a Holocaust survivor. When she asked about his family, he told her that he
was alone, and that he was the only one in his family who had survived.

As he spoke, she realized that there was no going back. She was
drawn to this man, to his life, and to his story. They were, in a way, similar. They had both lost their families and were determined to make the
most of their new lives in freedom. From then on, the two were a team.

By 1956, the communist youth movement in the East was in full
swing, with millions of children being born purely into communism. In
Schwaneberg, young Heidi was taking her first steps as a budding com-
munist by joining the Young Pioneers. Oma tied a red kerchief around
Heidi's collar and Heidi set off, hand in hand with sisters ten-year-old
Tutti and eleven-year-old Helga, to her first meeting. Heidi quickly made

friends and watched to see how the older children behaved; she was excited to see what all the fuss was about and eager to follow their lead.

The meeting was called to order and all the children ceremoniously encircled the Pioneer flag, saluting with thumbs to their forehead, hand facing out. In a perfect example of youthful communist discipline, a young girl several years older than Heidi corrected Heidi's salute, straightening her elbow, telling her to stand up straighter, and commanding Heidi to bark out the motto "Always ready" with real conviction and spirit.

Heidi took the Young Pioneer oath, promising to wear the red scarf, the flag of the Communist Party, with honor, vowing to cherish the Soviet Union, and to love and defend the "socialist fatherland." Over the next months, she learned cheery, carefully masked, propaganda-laden songs and watched films about happy Soviet life. She was proud to be selected for a bit part in a play depicting dedicated factory workers. But something had started bothering her.

At supper one evening, she asked Oma why everyone hated the West so much. Oma looked to Opa to answer.

"We need to protect ourselves against our enemies and some of them are in the West," he stated matter-of-factly, and began eating, signaling he was finished with the topic. Recalling no threatening characters on her trip to the West almost two years earlier, Heidi was confused. She became quiet, thinking it through. Several minutes later, she declared that she no longer wanted to attend Young Pioneer meetings. But by the next meeting, Opa had convinced her to return, assuring her it was for the best.

Then Heidi announced that she wanted to write a letter to Hanna to ask her a few questions. Oma gently talked her out of it, simply saying that it was not a good time to try to contact Hanna. Eventually Heidi would come to understand that there were some things best not written in a letter, especially to anyone in the West.

At around the same time Heidi became a Young Pioneer, Kai turned fourteen, and so it was time for him to join the FDJ. The Jugendweihe (youth initiation) ceremony, a tradition in Germany long before the Soviets took over, was akin to a religious confirmation, marking a young person's entry into adulthood. By the 1950s, however, the East German regime had co-opted the ceremony, using it to mark the dedication of one's life to communism.

The community hall was decorated with East German flags, flowers, and vibrant red banners that spanned the stage. The inductees' family members, who were required to be in attendance, sat in the audience dressed in what they once called their Sunday best. Kai took the stage with the other members of his class.

Local leaders gave solemn speeches about the importance of the ceremony as they called on the young socialists to strive to become loyal East German citizens.

The attending official turned to the inductees with serious bearing:

Are you prepared to use all your strength, to fight for peace
with all those who love it, and to defend it to the last breath?

Are you willing to fight side by side with us for a socialist
order of society . . . to fight together with all patriots for a
united, peace-loving, democratic and independent Germany?

"Yes, we pledge!" the young teenagers shouted, the sound echoing throughout the room until it fell sharply silent.

As the students stood on the stage, Oma watched her sweet and gentle sixth child nervously reciting the oath and wondered what life would hold for him as a man in communist society.

*I*n the West, Hanna and Eddie became almost inseparable, spending most evenings and weekends together. He introduced her to the American military and surviving German Jewish communities in Heidelberg. The two enjoyed exploring the city every chance they got, and began to take short trips to neighboring European countries.

Eddie had a voracious thirst for knowledge; he wanted to go everywhere and see everything. She loved being with him but his boundless energy and passion for adventure began to exhaust her. He dropped her off almost every night at around eleven o'clock, but after a while, with full-time work and her twice-weekly Italian classes, she suggested they go out only once or twice a week. One weekend, Hanna suggested they take the entire weekend off and catch up on sleep.

"You can sleep when you're dead," Eddie told her. "We're going to Paris this weekend."

Eddie was completely smitten with Hanna. She was a beauty but didn't seem to know it. She was independent, intelligent, uncomplicated. Hanna saw in Eddie an unusual zest for life. Despite what he had been through, he was always upbeat and had a great sense of humor. And he had that sparkle in his eye.

By December, the young East German runaway and the young American lieutenant were engaged.

*A*t home one evening, Opa confided to Oma that his situation at school was worsening, and that he was having a hard time tolerating the authorities' increasingly twisted demands. No longer able to stomach the regime's attempts to turn education into a propaganda tool, he was finding it difficult to hide his growing unhappiness. He was particularly appalled that he was now required to show favoritism to students whose parents were Party officials, or who were themselves leaders in the youth movement. Opa told Oma that this kind of cronyism would one day doom the system. At school Opa made no mention of his burden but kept up the façade. Now, at his Saturday-night card games, he had to fight harder against the impulse to speak his mind and share his growing displeasure with the regime.

One day Heidi's teacher addressed the class: "What do we think about people who abandon East Germany?" Not waiting for an answer, she continued. By "leaving," she explained, they had turned their backs on their country and their fellow citizens and this made them traitors.

"Traitors," she said, "are criminals . . . the absolute *worst* kinds of criminals."

Heidi listened intently, becoming increasingly disturbed as she envisioned Hanna's face. Having heard the same message from her Young Pioneer leaders, she wondered, was there something her parents weren't telling her? Were they trying to cover up for their daughter's delinquent character? Perhaps her teachers and youth leaders knew more than her parents did. Perhaps the beautiful things that she had believed all along about Hanna were just a misguided fantasy.

Suddenly her chest felt heavy and she had a sinking feeling. Tears welled up in her eyes and she looked around the room wondering if anyone knew that she had a sister who was a criminal.

Communism spread throughout the world. Mao Zedong took control of China and, together with communist North Korea, fought a bloody ideological war against the West on the Korean Peninsula. North Vietnam, formerly part of French Indochina, became communist in 1954, unleashing fears that the nations of Asia would fall like dominoes to Soviet ideology. In response, a wave of anticommunist hysteria gripped the United States when Senator Joseph McCarthy sparked a series of paranoid witch hunts to unmask suspected communists and communist sympathizers in America. In a first for the United States, two American civilians, Julius and Ethel Rosenberg, were executed for espionage after being convicted of providing the Soviets with nuclear secrets that helped them build their first atomic bomb.

In the fall of 1956, three years after the East German workers' rebellion, the people of Hungary took to the streets in an uprising against the Soviet-backed government to demand change. Moscow called in the Red Army, and thousands of Hungarians were slaughtered as the Soviets brutally crushed the rebellion. Once again, fearing war, the West had few options other than to stand idly by.

Days later, Soviet leader Nikita Khrushchev, in part citing the failed uprising, chided a group of Western diplomats when he lashed out at the West, threatening, "We will bury you!" The next year, the Soviets took the competition for ideological predominance into outer space as they launched Sputnik 1, the world's first artificial satellite. The same technology could also enable the Soviet Union to launch a missile that could reach North America. In addition to the nuclear arms race, now the space race was on.

The authorities in East Germany took every measure possible to dominate the population.

Valuing only those they could manipulate, the regime worked

ceaselessly to rid society of anyone who stood in the way of, or could
potentially undermine, its goals. Among them were intellectuals who
were often inclined to challenge the system, and those who did were
demoted, marginalized, or removed from society, ending up in reeducation camps or in prison.

Opa continued to appear to back the regime. At Party meetings, he
was a vocal supporter when he thought he needed to be. But as a man
who placed a great deal of importance on pure, objective education,
he was appalled when he was forced to distribute pamphlets such as
this one that attacked and belittled the intellectual segment of society:

You have been educated, gone to the university.
You have your Ph.D.
You are a gifted artist, a teacher, a technician, a major writer, a
   scientist.
No one needs you!
The work of half a lifetime,
All the sacrifice, all the learning is in vain!
Do you think you are better than the rest
Because you are educated,
Because you are an intellectual
And speak in elevated ways?
You are like stale beer!
There is no need for you any longer.

*O*pa found it harder and harder to cope with the way the Party
twisted information to feed its cause. By now, on most nights after
supper, instead of plunging into Marxist theory, he retreated to his

study to look over his favorite books of poetry in an effort to seek a break from it all. School officials and the authorities noticed a change in Opa's pattern of behavior and that the passion he once displayed was no longer evident when teaching communist doctrine.

One day, a couple of policemen, sent by someone in the local government who apparently thought Opa needed to be taught a lesson, paid him a visit. They walked into the house and carted away armfuls of his prized books on history, philosophy, and art.

*M*ore and more Oma took to her garden alone. It was a tranquil spot where she could retreat from the tension of her everyday environment. Her garden began to blossom with vegetables, berries, and colorful flowers that she cultivated with the greatest, most tender loving care.

In Seebenau, Kallehn was coping as best he could. No longer smoking Lucky Strikes, he now puffed on hand-rolled cigarettes made of poor-quality Russian *machorka* tobacco. Although forced to turn over his crops to the local agricultural committee, he was relieved that his farm had not yet been collectivized and he was still able to work his family's land. Ama Marit worried that the final surrendering of his farm to the regime would mark the beginning of the end for Kallehn.

Ten years in, the regime had taken control of every aspect of society. With increasing secret police control and a calculated campaign to strip citizens of their freedoms, the people of East Germany had little choice but to acquiesce.

10

THE FUR COAT
LAST MEETING
(1958–1959)

Let parents bequeath to their children not riches,
but the spirit of reverence.
—*Plato*

*I*n Heidelberg, Hanna worked at her job and in her spare time traveled in Europe with Eddie as often as she could. She sent postcards to the family in the East from every location, but became increasingly concerned when her correspondence went unanswered. Unbeknownst to her or the family, most of her letters and all of her postcards from her travels in France, the Netherlands, and England were confiscated on their way into the country. Similarly, the family's letters were being intercepted on their way out. The rare letter that had reached Hanna arrived opened with no effort to re-paste the envelope, was missing pages, and lacked any real information about the family.

But then one day, a postcard from Italy made it to Schwaneberg.

Oma was elated and celebrated by happily placing it up on the living room mantel for the whole family to enjoy. When Opa came home, he took it down and threw it into a drawer, berating Oma for her indiscretion, telling her she was foolish for displaying it where the children could linger over it and visitors who might be informants could see it.

Later, Heidi, now eight years old, retrieved the postcard and put it up in her room, where she ogled over its luminous salmon-colored sky and gondolas floating on the crystalline waters of Venice. Just as Opa feared, the more Heidi looked at the image, the more she became intrigued with the idea of the outside world. In school, teachers seldom spoke about what lay beyond the borders of the country, and now with Hanna's captivating postcard from Italy stretching the boundaries of her imagination, foreign countries became a curiosity, often taking on the allure of fairy tales.

One day in school, as the class studied the territories of Eastern Europe, the teacher directed the children to observe the map of East Germany. They discussed the major cities and towns, looked over the contours of the countryside, the hills and valleys and outlying areas stretching all the way to the border. Heidi noticed that, while all the Eastern Bloc countries—Poland, Bulgaria, Czechoslovakia, and the rest—were depicted in great detail, West Germany and the rest of Western Europe, in complete contrast, were shown only by a perimeter outline around a vacant gray mass, as if there were nothing there at all and the area was barren of any geographical features whatsoever. Forbidden to be curious about the West, on this day Heidi braved a question: "What's on the other side? Why is West Germany just a big blank?"

Glaring at Heidi, the teacher responded to the class as a whole, explaining that West Germany was a dangerous, desolate wasteland, a vast chasm of darkness and of the unknown, filled with violent convicts, hoodlums, and people who preyed on others. Staring coldly at Heidi and the other children, she added, "We don't really even want to know what's over there. We are lucky that we live in the East and that we have barriers to protect us from the evil that lies in the West."

Though it had been four years since her trip to the West, and her memories were fading, Heidi still recalled her time with Hanna in the West with great fondness. She found herself increasingly conflicted between the memories she held dear and the frightening things that her teachers, youth leaders, even her father insisted were true. By the time she turned nine, she no longer wondered where the truth really lay and had decided that, though there might indeed be depravity in the West, there must also be great good. Hanna became a symbol of the good that Heidi believed the outside world had to offer.

After working for a few years for the U.S. Army, Hanna had amassed a small savings that she wanted to send to the family in the East. Assuming West German deutschmarks would be intercepted, she took a risk to buy Oma a gift instead, then boxed it and taped it tightly shut. Several months earlier she had attempted a test run, sending smaller parcels filled with sweets and trinkets from her travels, but the family had not received them. Eager to make contact, she threw caution to the wind, posted the package, and hoped for the best.

A month or so later the package arrived in Schwaneberg. Whether by oversight, a clerk's mistake, or a simple act of decency, the big cardboard box arrived partially opened but with contents intact. The children squealed with delight, gathering round as Oma opened it. With their help, she pulled out the bundle from inside. Then she collapsed on the kitchen chair. It was a fur coat.

Oma had worn the same threadbare woolen coat for as long as anyone could remember. Now thinned with age, it was no longer protective against the freezing German winters. Though the "fur" was actually a horse-hair coat, having been bought from a colleague's aging mother, with monthly payments made over a two-year period, Hanna hoped not only that the gift to Oma would warm her throughout the winter, but that it would also confirm to her parents that she was doing all right in the West.

Opa called the package's delivery a miracle. At first he forbade Oma to wear the coat, knowing it would call attention, but soon he relented, allowing her to wear it on only the coldest days, after conceding that he himself had no money for, or access to, such a luxury.

*I*n the spring of 1958, ten years after Hanna's escape, Oma was thrilled to receive a letter in which Hanna wrote that she was getting married. Eddie, her fiancé, she wrote, was a Holocaust survivor who, after the war, had emigrated to America, joined the military, and returned to Germany as a U.S. Army officer. Though she assumed the authorities would not allow it, Hanna invited the family to the wedding.

Oma immediately put in an application to travel a second time to Heidelberg, hoping that the authorities would reconsider their decision not to let her travel again and show leniency in light of the special occasion. But they did not, and the request was rejected.

*I*n Heidelberg, Eddie's military superiors tried to persuade him to call off the wedding. Knowing how much it meant to the young lieutenant and survivor who had staked his entire future on becoming a U.S. citizen and an army officer, they told him he risked perma-

In the West, Hanna and Eddie marry in Heidelberg in August 1958.
The family in the East is not permitted to attend.

nent damage to his hard-earned career in intelligence if he married a former East German citizen, especially one whose family was still in the communist East. Eddie was livid, and responded, "You can't tell me who I can and cannot marry." With that he threatened to resign his commission over the matter. The commanding general interceded and, after a series of security interviews and an investigation that fully vetted Hanna, her U.S. citizenship papers were expedited. In the summer of 1958, Hanna and Eddie married.

At the U.S. Army chapel in Heidelberg, Eddie wore his blue army dress uniform and Hanna wore a simple white, knee-length chiffon shift and a pearl necklace. With no family in attendance, Eddie's boss and an office colleague were the witnesses. The rabbi said a prayer and Eddie stepped on the glass.

At the conclusion of the brief ceremony, they walked downstairs to a surprise reception. Nearly a hundred people had gathered to wish the newlyweds well, including the entire American and German Jewish communities of Heidelberg and Eddie's entire chain of command, including the G2, the general in charge of intelligence for U.S. Army Europe.

*I*n Schwaneberg, Oma was crushed that she was barred from attending the ceremony. *What has become of the world when a mother cannot even see her child on her wedding day?* she thought.

Opa, just as pained, took to pen to write this letter:

My dear, dear Eddie, my dear son. You both have decided to get married and begin your life together in love, loyalty, and truth. From Hanna's letter, I learned about the cruelty that you and your family endured and, because of your terrible fate, you are even dearer to my heart. I welcome you, with all my heart, into our family.

All our children have, through generations, been endowed with good health in mind and body, ethics, morals, and a good heart. I wish you both to continue this heritage for generations to come. If, at any time, you should need me, I shall welcome you with open arms. It is my deepest desire to do good deeds and I welcome you into our family.

Hanna and Eddie honeymooned in Brussels, Belgium. While there, they attended the 1958 World's Fair, where Eddie had been tasked with an intelligence collection mission. The fair was a huge exposition that showcased various countries' cultures, technical innovations, and scientific advances.

A family dinner in Schwaneberg, East Germany. Hanna's portrait is on the wall in Oma's direct line of sight. (*Clockwise from left:*) Heidi (*looking at camera*), Manni and his wife, Roland and his wife, Opa, Tiele (*serving*), Oma, Tutti, and Helga.

Wearing a business suit, Eddie, with his new wife, walked around the fairgrounds, eventually making their way to the Soviet Pavilion. While Hanna looked over a Central Asian costume display, Eddie maneuvered toward a mannequin dressed in the latest Soviet Army field uniform. U.S. intelligence had indications that the Soviets had developed a new fabric capable of defeating U.S. ground surveillance radars, but they needed proof. Armed with a pair of small scissors, Eddie waited for the perfect moment, then cut a corner swatch off the bottom of the uniform coat. He returned to Hanna, who was oblivious to what he had just done.

A few months later Hanna became an American citizen. Eddie became the first American liaison officer to the German BND,

Bundesnachrichtendienst, West Germany's equivalent to the American CIA. There he worked alongside Germans, some of whom had served in the SS and Gestapo during World War II. As Germany's main intelligence arm, the BND concentrated most of all on collecting intelligence on East Germany, running a network of agents throughout the East who consolidated information on every aspect of East German life.

*F*our months after the wedding, the authorities in the East suddenly had a change of heart and, to the family's surprise and great delight, granted Oma and Opa permission to travel to Heidelberg.

Opa was called to the police station. This time, in addition to the usual police authorities, there was a man who identified himself as being from the Ministry of State Security, the Stasi.

"Is it true your new son-in-law is an American army officer?"

"Yes," Opa replied.

"Is it true that he is an American army *intelligence* officer?"

"I don't know," said Opa, realizing instantly that any shred of truth in this news would further complicate things for the family.

"Well he is," the Stasi man informed him, "and that makes him an enemy to our country." Then he resolutely stamped some papers and said, "I am giving you two wonderful days in beautiful Heidelberg with your daughter and new son-in-law. Learn what you can about your son-in-law, about his work and his access to classified materials." And with that Opa was dismissed.

*O*n a cold December morning, the train left the station in the East and headed southwest. By evening it reached Heidelberg. Hanna easily spotted Opa, with his large frame, broad shoulders, and great

height, towering over the other passengers on the platform. He helped
Oma, who was wearing her fur coat, to descend from the train.

Oma beamed when she saw Hanna and the two fell into each
other's arms. Releasing their embrace only to look each other over,
Hanna saw that her mother's hair had begun to fade to a pearly gray.
With a wide smile, Oma pulled Hanna close.

Unsure about to how to greet his new son-in-law, an American
military officer, a man who represented the epitome of what the
Soviet Union and East Germany considered their most ardent enemy,
Opa looked haltingly at Eddie until Eddie approached him. The two
shook hands: the tall East German and onetime soldier in the Third
Reich meeting his new son-in-law, an Auschwitz and Buchenwald
survivor and now a U.S. Army intelligence officer.

Then Hanna turned to face Opa. The last time they had seen each
other was ten years earlier, when she had set off to register for teachers' college and had never returned. He was terribly irate then and for
years afterward had felt betrayed by her. He had become conflicted,
but over time his feelings of anger had given way to a hollow emptiness. Now he just missed his daughter.

Hanna stood before Opa, wondering if his disappointment in her
had abated. She would have understood if it had not. He stood there
in silence, towering over her, and for the first time in her life, she saw
a wounded expression in his eyes. She went to embrace him and the
two walked slowly and silently arm in arm to the car.

As they rode to Eddie's U.S. Army officer quarters, an apartment
in Patrick Henry Village on the base, Opa shed his insecurities and
seemed to come alive as he looked over every detail in Eddie's new
light blue and white Chevrolet.

With childlike delight, he pored over the interior, smoothing his hand over the leather seating, caressing the armrest, turning the crank handle to open and then close the window, stretching his neck to see the configuration of the dashboard panel, becoming enamored of everything about the car, especially since cars were a rarity in East Germany. At the apartment, Opa was fascinated especially with the refrigerator. In Schwaneberg, it was blocks of ice from the door-to-door iceman and a wooden box that kept their perishables cold.

That first evening, Oma brought out homemade cards, drawings, and notes that each of the children back home had written congratulating the couple on their marriage, welcoming Eddie to the family, and sharing a bit about their lives.

Hanna was most excited to hear news from Roland, but his note was short. Pained not to be able to associate with his sister who had defected and her new American husband, but realizing the risk that any contact could have for his rising career and his family's livelihood, he wrote only that he truly hoped Hanna was well. Sadness washed over Hanna as she read her beloved brother's all-too-brief message, and she silently cursed the regime that had forced them to grow apart, the hurt made even more acute when she saw the photograph of him standing with his wife of eight years. He had filled out and was still handsome at thirty-two.

Sweet messages and photographs followed. There was sixteen-year-old Kai, dressed in his FDJ uniform, his boyhood platinum hair gone brown; his face had thinned but he was still recognizable. Klemens had become a teacher, married, had a child. Manni, a conscript in his NVA military uniform, had also married. The little girls, Helga and Tutti, who were two and three when Hanna left, had grown up, were now almost teenagers, and Tiele, a young woman, was engaged to be married.

Finally, at nine years old, there was Heidi, with her long, dark braids, looking just as inquisitive in her Young Pioneer photo as she had four years earlier in Heidelberg. Heidi had drawn a picture of two girls holding hands in front of the Heidelberg Castle and wrote how she missed Hanna and hoped to see her again soon.

*E*ddie showered his new mother- and father-in-law with affection. They spent the next day driving in and around the city and taking in the sights. Hanna delighted especially in showing Opa the castle.

As they walked the grounds of the fortress, he carefully examined its architecture. In the interior corridor, he took his time to indulge in the details of the intricately designed wooden ceilings and statues of nobility of the royal court. He ruminated on the castle's history and looked over every detail with professorial focus, finally musing that, despite war and fire, the castle had survived complete destruction. Hanna looked back at him and it was as if she were a young girl again, in Opa's classroom or in his study.

That evening, Opa played the piano, Eddie's wedding gift to Hanna. He played old German folk songs from his youth, and waltzes and classics, Bach's Minuet in G, Beethoven's Ode to Joy, as they all talked and sang into the evening, drinking Eddie's homemade *erdbeerebowle*, a potent punch made with champagne and cognac-sugar soaked strawberries. The evening grew more festive as it went on, the four of them becoming completely at ease in one another's company, realizing that every moment together was precious. Oma smiled her gentle smile nearly the entire time.

*D*uring the two-day visit, Opa became impressed with Eddie, not only for his comfortable U.S. Army officer lifestyle, but more important, for the depth of his character, and especially his remarkable ability in having rebounded from his tragic past. Opa was drawn to Eddie's joie de vivre and positive outlook for the future.

By the morning of the second day, Opa had clearly relaxed. He was slow to come to breakfast, which was very out of character for him; he was always an early riser. In the kitchen, Eddie reported to Hanna that, during the night, someone had gone into the refrigerator and finished off the remaining *erdbeerebowle*. Hanna laughed, happy to know that her father had made himself at home.

They spent the next afternoon walking in downtown Heidelberg. When the sun began to set and evening rolled in, it became colder, so they sat bundled up and close packed together to keep warm on a bench on the Philosophers' Walk. They lingered there, talking in the cold night air under a full moon as they gazed out over the city and at the castle, which was illuminated in a lustrous golden glow.

Back at the apartment, Hanna prepared dinner, a succulent beef *rouladen* from Oma's recipe. Eddie served French wine. Dessert was cognac crème followed by even more French wine. Opa felt like a king.

During the course of their last evening, Hanna finally had enough courage to ask how things were at home. Oma said, "They're fine. We're getting used to it."

Opa interjected, "Actually they're not fine and it's getting worse every day." Wanting to avoid distressing talk, Oma reminded him that they had only one evening left to enjoy one another's company. As the evening drew on, not wanting to dilute the richness of the time they had left with empty conversation and idle chitchat, they found themselves sitting for long stretches in silence.

*T*he next day, at the train station, Opa seemed genuinely happy knowing his daughter had established a fine life for herself. Despite

Oma, in her fur, with Opa and Hanna in Heidelberg in December 1958

mixed emotions over the last decade, he had finally made peace, perhaps even taking solace in the fact that he had likely played a role in
enabling his daughter to follow her dreams.

Oma too seemed comforted. She pulled Hanna close, looked into
her eyes, and said, "You made the right decision. I am happy you are
free." And with that, mother and daughter took a long time to let
each other go. When Hanna embraced Opa, she couldn't help but
feel that he too was trying to show that he harbored no grudge and
was in fact finally prepared to release her.

They delayed the departure as long as they could, finally boarding at the whistle. When the train started to move eastward, Opa,
pulling himself up to his full height, stood motionless gazing out the
window at Hanna and Eddie with a melancholy expression. Oma
waved, smiling through her tears.

It was the last time Hanna would ever see her parents.

———

Upon their return from Heidelberg, the Stasi summoned Opa to ask him about his new son-in-law, to which Opa simply replied that he could not get him to talk. The Stasi man looked back in a dead stare, then shook his head and added another black mark of disloyalty to Opa's growing list.

Though Oma applied several times after that day to visit Hanna again, all subsequent requests for travel were denied. Oma and Opa's failure to help the Stasi obtain information about or create a contact in American intelligence had made it clear that Opa was not willing to play the game.

PART THREE

11

"A WALL WILL KEEP THE ENEMY OUT"
A WALL TO KEEP THE PEOPLE IN
(1960–1961)

No one has any intention of building a wall.
—*East German leader Walter Ulbricht*

*I*n Schwaneberg, the Stasi knew Opa had lied to them. They made note of his disobedience but it didn't stop them from pursuing Eddie in Heidelberg. The very next year, they dispatched Opa's nephew, Edgar, to spy on Eddie and try to recruit him to work for the Stasi. But after Eddie fed Edgar a few beers, he started to cry and confessed to his spy mission.

"If I don't get the information they want, they'll cause big problems for me and my girlfriend," he said, utterly distressed. "We have a child. I don't know what to do."

After a while, he finally looked up and said, "I can't go back."

Hanna and Eddie dropped him off at the Heidelberg train station.

A few months later, they got a postcard from Edgar, who had made his way to Paraguay.

*I*n Seebenau, inasmuch as he was still able, Kallehn, now in his seventies, adjusted to communist life, though in truth, he suffered in silence. Few in the family had been able to visit him since the enactment of the now five-year-old rule that had shut off border towns to all but the residents of those areas. To make matters worse, as he grew more fragile with age, Kallehn could no longer travel to see his family in Schwaneberg, which virtually banished him to a life of isolation and internal exile.

By 1960, the year of the "socialist spring," the East German government collectivized the remaining independently managed farms. Farmers who had avoided giving up their land were now forced to do so. Kallehn had resisted, holding on as long as he could, but, not wanting to go to prison and lose everything, he finally gave up and handed over his family's land. Exhausted and dispirited from it all, in late 1960, Kallehn passed away. For a while, Ama Marit remained in the farmhouse, but then moved in with her daughter Frieda, who still lived in her little house on the edge of the border.

By the 1960s, tensions between the two superpowers were escalating. Cuba had aligned with the Soviet Union and adopted Marxist ideology, which resulted in the United States cutting ties with that neighboring island. At a meeting at the United Nations, a Western dignitary, the head of the Filipino delegation, made a speech in which he stated that the peoples of Eastern Europe were being deprived of their freedom, and that their countries had been "swallowed up" by the Soviet Union. Soviet secretary Khrushchev responded by bolting to the podium, banging his shoe, and calling the diplomat "a jerk, a stooge," and a "toady of American imperialism."

*I*n May 1960, the Soviets shot down a U.S. spy plane that was flying in Soviet airspace. The pilot, Francis Gary Powers, spent a year in a Soviet prison. The United States initially denied the downed plane was used for spying, but after Khrushchev produced evidence, a humiliated President Eisenhower had to publicly admit the plane was on an intelligence mission. The incident was a key moment in the Cold War and is credited in large part with the collapse of arms control talks that were taking place between the United States and the Soviet Union in Paris.

The two rivals continued to grow their nuclear stockpiles and by the early 1960s, the Soviet Union had joined the United States as a bona fide nuclear superpower.

*T*he space race continued. Following the successful Soviet launch of the first satellite, the Americans launched their own. But then the Soviets upped the ante. Less than one month later, cosmonaut Yuri Gagarin became the first man in space, a victory for Moscow and communism. The United States followed by sending astronaut Alan Shepard into space, to which the Soviets responded by sending the first woman into orbit and then raised the bar higher by conducting the first spacewalk. That was followed by an American spacewalk; not long after that, John Glenn became the first man to orbit the earth.

*I*n East Germany, escapes continued. By 1960 an estimated three and a half million East Germans, nearly one-sixth of the population, had fled.

As people found ways to get out, the East German authorities feared that West German agents were making their way in. Thus the Stasi began arresting hundreds of East German citizens and charging them as foreign spies. With a heavily fortified border separating East from West Germany now a huge deterrent, and the authorities still unable to control access into West Berlin, Berlin remained the best hope for those who wanted to escape. By now the Marienfelde Refugee Center in West Berlin was processing some two thousand East Germans a day. With East Germany's population dwindling and its economy on the brink of collapse, the time had finally come, Ulbricht decided, for something drastic to be done to stop the outflow once and for all.

There was no word at all from East Germany. In Heidelberg, Hanna had not heard from her family in more than two years, since her parents had visited. In Schwaneberg, the authorities prevented the family's letters from leaving the country, likely due at least in part to Oma and Opa's failed mission.

With no news from her family, Hanna resigned herself to scouring West German newspapers and tuning in to television and radio to learn what she could about what was happening inside East Germany. Information was sparse, however, and finding out any news about small villages like Schwaneberg was nearly impossible. For Hanna it was as if East Germany and her family were beginning to slip away from her and fade into a remote gray fog.

Hanna and Eddie had their first child, a boy, Albert. Then the U.S. Army posted Eddie back to America, where John F. Kennedy had just taken the oath of office as president. From the start, Kennedy was enormously popular among Americans, in no small part due to his tough stance on the Soviets.

Hanna was thrilled to begin her new life in America but was disheartened that the move would put a greater physical distance between her and her family, even though it no longer seemed to matter anymore whether she was one hundred or three thousand miles away.

They moved to the prairies of Kansas, where Eddie was assigned to Fort Riley. I was born in March 1961, as Hanna, my mother, started her new life in the Midwest as an American military wife, with now two babies.

That same month and into the spring, thousands of East Germans were still scrambling into West Berlin every day. Rumors circulated throughout the country that the regime was preparing to permanently seal off West Berlin by building a physical barrier around the city. Believing the window for escape might be closing for good, large numbers of East Germans rushed the border. In mid-June, Ulbricht, worried that the rumors were sparking a panic and an exodus that would be catastrophic to his economy, took to the airwaves to tell the citizens of East Germany that their fears were completely unfounded.

"No one," he assured them, "has any intention of building a wall."

But in the early-morning hours before dawn, on August 13, 1961, while Berliners slept, brigades of construction workers and some forty thousand East German troops set to work rolling out spools of barbed wire and erecting fencing to block off all access to West Berlin.

Armed East German soldiers stood at six-foot intervals, prepared to fire on any last-minute defectors, who, they had been warned, might include the construction workers and even fellow border guards themselves. Concerned that NATO might respond with force, Soviet tanks had taken up positions along the border, but when there was no response, East Germany's *Aktion Rose* (Operation Rose) went into full swing.

Deafening noise riddled the city as heavy machinery ripped up streets and knocked down buildings. Roads, subway systems, and rail lines were severed. By midmorning, thousands of West Berliners had amassed to watch the extraordinary scene unfold. Shocked at what they were witnessing, they stood watching, shaking their heads and verbally assailing the East German soldiers and construction workers who appeared unaffected by their taunts and quietly went about doing the work they had been ordered to do.

By noon, all routes into West Berlin, the last border that had offered any chance of escape, were effectively sealed for good.

*P*eople all around the world stopped in their tracks to see and hear the news, astounded by the eerie scene of a country locking its people in.

In her military quarters in Kansas, Hanna got up that morning and tended happily to her two babies. With the summer sun streaming through the window, she made herself a cup of coffee and switched on the television, preparing to start her day. Tuning in, she was surprised to see a live news broadcast from Berlin reporting that East Germany was being cut off from the West.

The scene showed East German workers toiling away on a massive construction project. Initially confused, my mother suddenly felt ill when she understood what she was seeing. Watching the workers stacking concrete blocks, armed guards in the background, she was heartbroken as she realized that the regime had taken final desperate measures to break from the West. A wall would seal off the country and pull the people of East Germany even further into isolation and, she knew, would take her family with it.

*W*ithin twenty-four hours, the Berlin Wall cut through the heart of the city. By the next day, the Brandenburg Gate stood trapped inside the East, only fifty paces from freedom. The very symbol of German unity, that day the majestic Brandenburg Gate would ironically become the foremost symbol of a divided Germany.

*O*ver the next days, the pace of construction intensified as the first generation of the Berlin Wall was erected.

Made up of barbed wire and concrete blocks, it would stretch over a hundred miles, and completely encircle the island city of West Berlin, cutting it off from the rest of East Germany. Construction severed streets, bore through neighborhoods, even sliced cemeteries in two. East Germans who happened to be in the West at the time of construction were simply cut off from their loved ones on the other side, families separated in an instant.

Hanna watched television images of distraught family members waving to one another from opposite sides of the border until East German police shooed away those in the East.

More tragic scenes followed: desperate people jumping from East Berlin apartment buildings onto the West Berlin pavement below, West Berliners on the other side waiting to catch them and assist in their escape. Those images were followed by scenes of workers calmly boarding up windows and forcing people to relocate farther into East Germany.

The world continued to watch as hundreds of watchtowers sprouted along the Wall. At the same time, the Inner German border between

East and West Germany was fortified with a greater number of, and more complex, obstacles, which eventually included mines and automatic spring-loaded guns.

*M*oscow watched to see how the young, untested leader in Washington would react to the sealing of West Berlin. President Kennedy, mindful of tensions escalating to dangerous levels, vehemently objected to the building of the Wall, but not wanting to risk World War III over it, conceded that "a wall is a hell of a lot better than a war." As long as Western rights in Berlin were not threatened, he said, the United States would not interfere.

Of the East German people, West German chancellor Konrad Adenauer pledged, "they are and remain our German brothers and sisters." West Berlin mayor Willy Brandt, who went to witness the building of the Wall up close, called it "the Wall of Shame."

Moscow was relieved that the West took no action to challenge the situation. With the pressure of retaliation off, Ulbricht turned to face the people of East Germany. He told them that the Wall was an "antifascist protection barrier," and that it had been designed to keep them safe from attacks from the West. Without such a looming threat, he said, the state could start making real progress. The next day, the East German newspaper *Neues Deutschland* ran the headline, "Measures to Protect the Peace and Security of the German Democratic Republic in Force."

*T*he first known leap to freedom over the Berlin Wall came on August 15, just two days after workers began building, when Conrad Schumann, a nineteen-year-old East German border guard who had volunteered for duty in Berlin stood at his post, rifle at the ready,

with orders to keep his countrymen from escaping. As West Berliners called "come over here," Schumann jumped the barbed wire and
bolted into West Berlin, where he was whisked off by a waiting West
Berlin police car.

Nine days after construction began, news rang out about the first
known casualty at the Berlin Wall. Fifty-nine-year-old Ida Siekmann
died after trying to leap onto a West Berlin street from her third-floor
apartment window before the building was condemned to make way
for more border construction. Two days later, Günter Litfin died after
being shot in the back of the head by an East German border guard
as he attempted to swim across Berlin's Spree River to the West. In
the East, the regime conducted a smear campaign, slandering Litfin
as a hooligan with a criminal past long before he tried to flee. West
German news condemned the murder as "brutal cold-bloodedness."

Over the next years, the Wall would undergo constant upgrades in
an attempt to make it increasingly impossible to penetrate. Border
officials carefully studied every spot where anyone had escaped and
corrected deficiencies in the structure so others could not escape in
the same way.

What began in the early hours of that warm August morning as a
simple barbed-wire fence would soon evolve into a twelve-foot-high,
one-to-three-foot-thick concrete structure with a rounded top to prevent grasping and scaling its heights. Wire mesh and electric signal
fencing and more rolls of barbed wire would eventually be installed,
along with various electric alarms, searchlights, trenches, and, all
along the border, the death strip, a hundred-yard-wide gauntlet of
meticulously raked sand to make it easy to spot the footprints of
escapees. At night, floodlights and searchlights constantly scoured

for signs of life. Booby-trapped throughout with tripwires and anti-vehicle trenches, the Wall was a death trap that offered a clear field of fire for armed guards who were posted in some three hundred watchtowers around the perimeter, with orders to ensure no one got over it alive.

*J*ust months after Wall construction began, an incident occurred at Checkpoint Charlie, the Allied crossing point from West into East Berlin, when East German border guards denied a U.S. diplomat unhindered access into the Soviet sector, East Berlin, in direct violation of the Four Power agreement.

In response, the U.S. Army's Berlin Command moved a column of ten M48 A1 Patton tanks right up to the edge of Checkpoint Charlie, facing the Soviet sector. American diplomats, escorted by U.S. military police, pushed their way into the East on foot. The next day Soviet tanks moved ten T-55 tanks into opposing positions on the east side of the checkpoint. With the attention of the world once again locked on Berlin, American and Soviet tanks, barrels trained on one another only yards apart, menacingly faced one another down. Nerves on edge, many wondered if the world was on the brink of World War III.

Thanks to back-channel diplomacy between Khrushchev and Kennedy, twenty-four hours after the standoff at Checkpoint Charlie began, tensions abated when the Soviets withdrew one tank. The United States in turn removed one tank. The pullout continued slowly, one tank at a time, until all had retreated. The world breathed a collective sigh of relief, and Allied access into East Berlin resumed.

From her distant perch in Kansas, my mother had no way of knowing that her now eight-month-old baby would one day be part of a team that would regularly cross through Checkpoint Charlie to

West Berliners peer through the Berlin Wall into the East
near Checkpoint Charlie.

exercise those American rights to access, and to run intelligence collection missions, in the Soviet sector of Berlin.

In Schwaneberg, the family learned about the building of the Wall from Ulbricht's radio address to the people of East Germany. Despite their leader's explanation that the structure was erected to protect them, Oma and Opa knew it was being built to lock them in and cut off connection with the West.

The building of the Berlin Wall marked a turning point for Oma. She now feared the family was severed from Hanna for good. With no contact in almost three years, she did not even know that Hanna had moved to the United States, or that she had had two children.

*P*arty bosses, teachers, and youth leaders throughout East Germany were ordered to spread Ulbricht's message that the Wall had been a necessary measure to keep them safe. In Schwaneberg, Heidi's teacher made her point by asking if anyone in the class wanted the evil forces of the West to destroy the country.

"We have to make sure East Germany is locked up tight against those who want to harm us," she said.

The children looked at one another, their concerns falling away to pride that their leadership had taken such bold measures to stand up to its enemies and protect its citizens.

*W*ith all avenues of travel out of the country sealed, the regime turned back to manipulating the population. Oma braced for the further tightening of controls that threatened her children. Now even more compelled to give her family a soft place to fall when things got rough, she did the only thing within her power to protect them—she built a barrier of her own and instituted a rule of family solidarity at all costs.

The safe haven she had begun to create the day the Soviets stepped foot in Schwaneberg, to shelter her family from the suffocation of the regime, now had a name. She declared the Family Wall a sanctuary, a refuge where the family would preserve their souls by keeping the good in and the bad out. The children followed Oma's lead and the concept took hold.

Inside the Family Wall, the children let down their guard. As the fabric of East German society began to fray under the yoke of an Or-wellian climate of oppression, and families wondered whether or not they could trust their spouses, parents, or siblings, Oma demanded family trust and loyalty. Behind closed doors, to Opa, Oma insisted that they had to foster the idea of the Family Wall if they were to have any chance against a regime out to crush the spirit of its people.

Back in the village, however, no matter how much Oma tried, the Family Wall could not keep the authorities from invading their lives. They continued to harass Opa, whom they directed to work harder to encourage his students to passionately serve the regime. When the authorities began to detect apathy in him, they taunted him by insinuating he was losing his edge and that his days were numbered.

12

THE FAMILY WALL
OMA'S FAITH AND OPA'S DEFIANCE
(1962–1965)

Freedom has many difficulties. And democracy
is not perfect. But we have never had to put a
wall up to keep our people in.
—*President John F. Kennedy*

*E*ast Germany now had a profound image problem. The Wall had
added insult and injury to a reputation that was already dismal, so
the leadership assembled to figure out a way to deal with the thrash-
ing it was taking in the eyes of the free world.

Meanwhile, throughout the East, Party bosses praised the regime's decision to secure the borders against the enemies in the West.
In Schwaneberg, at a Party meeting, the authorities tested community leaders, asking for their opinions about the building of the Wall.
While the ardent new village mayor supported it outright and others
enthusiastically advocated for it, Opa nodded his approval but said
nothing; it all came off halfhearted and sorely unconvincing.

At school, Opa was compelled to back the Party's explanation for the Wall and none of his students challenged him. As with many other topics, the villagers did not discuss with one another the sealing off of East Germany. Some East Germans accepted the authorities' reason for the building of the Wall and simply went on with their lives.

Heidi and her sisters continued to divide their time between home, school, and youth activities. Helga and Tutti took Jugendweihe and moved up to the FDJ, while Heidi remained a Young Pioneer. After his service in the army, Manni became a teacher like Roland, Klemens, and Tiele. At the age of nineteen and just out of high school, Kai was drafted into the NVA to serve what by now had become an eighteen-month tour of compulsory military service for every young East German male.

*W*hile life under communism had now become routine for the family, Opa seemed to be the only one who could not manage to make peace with his reality. At home, his agitation began to show itself in uncensored outbursts. Roland, by now a mature man in his mid-thirties, found himself constantly warning Opa to be more discreet, if not for himself, then at least to consider the possible consequences for his family. Opa tried to heed his oldest son's advice. But he could not contain his anger the day he learned what the regime had in mind for his fourth son, his gentle and sweet-tempered Kai.

During training, Kai distinguished himself as a top performer, earning honors as a standout athlete and an expert rifle marksman. As a result, he was ordered to serve his country as a border guard at the Berlin Wall.

Opa was upset. Though he knew that Kai had no choice in the

matter, he did not want his son performing border duty where he
would be put in the position of having to shoot at someone trying to
escape. Oma spent much time in her garden that summer, thinking
about and praying for her boy.

Around the same time I was turning two years old in the United
States, Kai was put through a battery of tests to ensure he was politically reliable and capable of serving on the border. He donned his
olive gray border guard uniform and reported for duty. Preventing
escapes, his superiors told him, came with rewards, medals, and promotions. Failure to prevent escapes, on the other hand, would be
punished by a prison term, consequences for his family, and eroded
prospects for his future.

Because of their proximity to the West, guards had to be closely
watched and constantly evaluated. At every posting, Kai was paired
with a new guard, someone he had never worked with before. Required to serve in pairs, they could work together only once, minimiz-
ing the chance that they might form a bond, then conspire to escape,
or that they would agree not to fire on would-be escapees. It was their
sworn duty to report any hint of disloyalty in partner-guards, including signs that they were not committed to accomplishing the mission
to its utmost end.

Most guards followed the rules, but a few tested the system. Rüdiger Knechtel was sentenced to a year in prison for tossing over the
Wall a bottle that contained a note to the American Forces Network (AFN) in Berlin, in which he expressed his disillusionment
with life in East Germany and wrote the word *Schandmauer* (Wall of
Shame), then requested a song. Without mentioning his name, AFN
promptly dedicated a song only by saying, "Congratulations from the

other side," then posted the note on a bulletin board, where it was seen by a Stasi spy who reported the note to the Stasi.

The United States and the Soviet Union continued to wage proxy wars throughout the world, each side buttressing its capacity to conduct nuclear war. As things looked to be heating up, President Kennedy encouraged Americans to build underground bomb shelters to protect themselves in the event of a Soviet nuclear attack.

In February 1962, the Soviets released U.S. pilot Francis Gary Powers back into U.S. hands in exchange for a Soviet spy. The exchange took place over the Glienicke Bridge on the West Berlin–East German border, a location that would be made famous in the Cold War years for its Soviet–U.S. spy swaps.

That same year, the United States increased its involvement against the communist insurgency in South Vietnam, and in October the world was brought to the brink of nuclear war for thirteen days during the Cuban Missile Crisis, when the Soviets sent nuclear-equipped missiles capable of reaching the United States to Cuba. President Kennedy challenged the Soviet Union:

> It shall be the policy of this Nation to regard any nuclear
> missile launched from Cuba against any nation in the
> Western Hemisphere as an attack by the Soviet Union on the
> United States, requiring a full retaliatory response upon the
> Soviet Union.
>
> I call upon Chairman Khrushchev to halt and eliminate this
> clandestine, reckless, and provocative threat to world peace and

to stable relations between our two nations. I call upon him
further to abandon this course of world domination, and to
join in an historic effort to end the perilous arms race and to
transform the history of man. He has an opportunity now
to move the world back from the abyss of destruction.

By mid-1962, more than twenty people had been killed at the
Berlin Wall and some two hundred had been caught trying to escape
over it. On the first anniversary of the building of the Wall, the citizens of West Berlin paused for three minutes of silence for those who
had been shot or died falling from buildings or in the Spree River.
In the coming years many more would be murdered while trying to
make it to freedom.

One year after the Wall was built, Western media captured an incredible, heartbreaking failed escape on film. The world watched,
horrified, as Peter Fechter, an eighteen-year-old bricklayer from East
Berlin, attempted to scale the Wall near Checkpoint Charlie, and
was shot by border guards in plain view of onlookers from West
Berlin. Despite his pleas for help, he was given no medical assistance
as he lay bleeding at the base of the Wall, East German border guards
threatening to shoot those from the West attempting to intervene.
Livid West Berliners lashed out, screaming "Murderers!" as Fechter
slowly bled to death. An hour after he was shot, East German border
guards retrieved his lifeless body and carried him away.

Back in the village, Opa's superiors finally summoned him to address his waning passion for communism. He braced for another

round of denunciations. They told him to sit in a chair in the middle of the room. One by one, they chastised him, telling him that he lacked the required zeal for socialism and that his mediocre performance was an insult to the Party. As usual, Opa said nothing in his defense.

For the time being, he was allowed to remain in his role as headmaster, but now he knew it would not last. The state no longer had faith in him and began to seed the village with mistrust against him.

That summer, as she watched her father's mounting despair, Heidi turned thirteen years old. While Oma always seemed to find joy, even under their repressive circumstances, Opa did not, and now, more than ever, he seemed distant.

*W*ith the country on lockdown, the secret police intensified their control of the population, perfecting their methods of penetrating every aspect of a person's life, including reading their mail, listening in on conversations, gathering compromising details in an attempt to identify weaknesses that could be used to manipulate them, even threatening to expose family secrets of indiscretions or exploiting character flaws to the secret police's advantage.

The Stasi recruited more agents and perfected the dirty business of blackmail and bribery, offering promotions, special perks, or money, or holding career or educational advancement over people's heads to get them to comply. They rewarded those who helped them and punished those who didn't.

The informant program also grew, the Stasi playing people off one another in an attempt to control everyone; the program was a success because no one knew who could be trusted and who the informants were. By now there were citizen spies in every factory, social club, and youth group. Every school, apartment building, military unit, polit-

ical group, and sports organization was a potential pool of intrigue
and exploitation. Without knowing who in their midst was eaves-
dropping or gathering information to pass on to the secret police,
self-censorship became a way of survival.

Fewer and fewer challenged the state anymore, knowing that com-
plaints and comments against the regime or East German policies
were punishable. Any act of defiance could cause people to disappear
or destroy their own or their children's futures. So people tried to
stay off the Stasi radar and kept their heads low, hoping to live a quiet
life under whatever conditions they faced. Since there was no sign
that anything was going to change, people resigned themselves, made
peace with their situation, and built a life accordingly.

*F*ed by the continuing paranoia of Western influence, Ulbricht
further isolated his citizens when he demanded the regime double
its efforts to keep people from turning to Western media for news
and entertainment. The government imposed harsher punishments
and set FDJ youth to find and destroy or remove antennae that were
pointing westward, in a campaign called *Blitz contra NATO-sender,*
Blitz against NATO signals.

The world remained in shock at the imprisonment of the people of
East Germany. For those separated from their families, time just stood
still. Melancholy over the separation of the two Germanys lingered. In
the United States, Toni Fisher's haunting ballad "West of the Wall,"
about families and lovers separated by the barrier, became a Top 40 hit.

*F*or my mother, Hanna, after the Berlin Wall went up, East Germany and her family seemed to detach and pull away, disappearing

further and further behind the Iron Curtain, into a mysterious, remote world where she feared she could never reach them again.

In her home in Kansas, Hanna read a *Time* magazine article titled "East Germany: They Have Given Up Hope," which relayed a West German traveler's visit to his relatives in the East German city of Dresden. He said, "Never have I seen people who feel so alone, lost and abandoned. They have given up hope." Hanna dismissed the report, choosing instead to believe that her family was somehow getting by.

In fact, while the early years were spent struggling to rebuild from the war and adapting under the Russians, most were by now adjusting under East Germany's brand of authoritarian communism. Some embraced the system while others played the game just to survive. Many put up defenses aimed at maintaining their personal dignity, but most now realized that one could have a quiet, uncomplicated life provided one didn't expect much and didn't challenge the authorities.

At fourteen years old, Heidi was due to take Jugendweihe. Though she had been conditioned to look up to those who had taken the oath, unlike her classmates she no longer looked forward to the ceremony vowing loyalty to the regime. When her time came, Opa, warning that her future depended on it, pushed her to go, and so, among her peers and youth leaders and banners praising socialism, she too raised her right hand and dedicated herself to East Germany, then dispassionately joined the ranks of the FDJ.

Opa spent the winter feeling increasingly isolated in his thoughts, so when spring arrived, Oma tried to get him to reconnect to the outdoors and to nature, which he had always loved. They often worked together in the garden, she noting which flowers had bloomed, which

VOR EUCH DER SOZIALISMUS –
WELCH EIN ZIEL!

Heidi (*far right*) takes Jugendweihe and ponders her future under
communism.

plants needed more water, as he raked the soil and picked vegetables
for their next meal and for pickling. Together they made raspberry
jam and apple cider, took walks in the grassy fields, or wandered in the
forest picking mushrooms and poring over the flora. In the woods, he
often walked ahead or moved off in a different direction, preferring to
be alone for a time or wanting to find a place to sit by himself.

The whole family knew that Opa was walking on thin ice. Oma
felt that it was in his and the family's best interest for Opa to have
some joy infused into his life, so she invited the family to Schwaneberg to celebrate his birthday. Though Kai was still on border duty,
everyone else came: Roland, Klemens, Tiele, and Manni and their
families arrived with gifts of flowers, food, and homemade liqueur.
Teenagers Helga, Tutti, and Heidi buzzed around the kitchen helping Oma prepare soup and potatoes, and plate pickled vegetables

from the cellar. Though they had never really understood how Oma, with her meager rations, somehow managed to get enough to put a big spread on the table, she always did, even now as she set out homemade cakes, *Butterkuchen, Bienenstich,* and *Apfelkuchen.*

Roland stood before the family, holding his glass high to make a toast.

"Papa," he said, "today we gather to celebrate you on your birthday. And, as with every year on this very day, we also commemorate East Germany's Day of the Teacher. Papa, we all look up to you as our wonderful father and role model, and we are all proud to have followed in your footsteps as teachers. We know that this year has been a challenging one for you, but your family will always be here to support you. Many wishes for much health and many years of happiness. *Zum Wohl.*"

That afternoon, everyone seemed to drift back to a simpler time. Oma steered the conversation to more carefree days of the past. Memories flooded in. Roland, Klemens, Tiele, and Manni recalled winter in the little village of Trabitz, where they had lived before moving to Schwaneberg. In heavy snow, homemade sleds and wooden crates would be tied together and, with their teacher, Opa, at the helm, the children gleefully sailed down Kanterburg Hill, a couple of times sliding terrifyingly close to the frozen Saale River. Helga recalled how when a stork had built a nest on Herr Poppel's stable, it was Opa who forbade anyone to tear it down, and how after that day the bird became the village mascot and Opa was given the moniker the Stork Rescuer.

That evening Oma persuaded Opa to play some of his favorite folk songs on his beloved Schimmel piano. He did so and the children joined in, singing in harmony. After the sun went down, they built a bonfire outside and hovered around it. Looking at his children laughing in the glow of the fire, Opa smiled, and for a while he seemed to be all right.

Despite the building of the Berlin Wall, those still determined to get out would not be stopped. To have any chance at success now, though, they had to invent new ways to outsmart border guards. A few got out hidden in the trunks or tiny front engine compartments of cars authorized to cross into the West.

Then, in April 1963, Wolfgang Engels, a nineteen-year-old civilian employee of the East German Army, stole an armored personnel carrier from the base where he worked and crashed it into the Wall. When the vehicle failed to completely penetrate the concrete, Engels bolted, running on foot through the rubble toward the West while East German border guards fired on him. A West German policeman intervened, firing his weapon at the East German border guards. Shot twice while struggling to free himself from barbed wire, Engels was pulled to safety by West Berliners and survived. After this escape, an upgrade was made in which three-meter-high, three-ton wall sections were put in place to block any similar attempts.

In another escape, Rudolf Mueller, who had fled to West Berlin before the Wall went up, dug a seventy-two-foot underground tunnel from West to East Berlin to get his family out. As he moved his family into the tunnel, they were discovered by a twenty-one-year-old border guard who raised his weapon and ordered them to stop. Mueller shot and killed the guard before the family escaped to West Berlin.

In the East, the Stasi instructed citizen informants to find out who might be plotting to escape, and how and where they were planning to do it. To their border guards, the regime ordered, "Do not hesitate with the use of firearms, including when the border breakouts involve women and children."

*I*n June 1963, President Kennedy visited Berlin. Near the Brandenburg Gate, the leader of the free world climbed the steps of a viewing platform and looked over the Wall into the East at the obstacles and barriers meant to keep East Germans imprisoned. On the other side, armed guards stood looking back, all the while monitoring the death strip below.

Addressing a crowd of 120,000 outside West Berlin's city hall, the Rathaus Schöneberg, in a message meant to be a show of unwavering support for West Berliners and for their freedom, and carried on live television all over the world, Kennedy said, "There are many people in the world who don't understand the greatest issue between the free world and the communist world. Let them come to Berlin.

Freedom has many difficulties and democracy is not perfect, but we have never had to put a wall up to keep our people in to prevent them from leaving us. West Berliners and people around the world went wild as Kennedy concluded with these now-famous lines: "All free men, wherever they may live, are citizens of Berlin, and, therefore, as a free man, I take pride in the words 'Ich bin ein Berliner!'"

Kennedy's remarks inspired the Western world and gave a morale boost to the people of West Berlin in one of the most important moments for the United States during the Cold War. Five months later, in November 1963, Kennedy was shot and killed in Dallas by Lee Harvey Oswald, an American who had once defected to the Soviet Union.

———

News from the East was sparse for almost a decade. Teenage Heidi longed to be in touch with her sister, if only to share news about important family events she was sure Hanna would want to know: that Roland had been promoted to headmaster of his high school, and that he and his wife had a five-year-old boy; that Manni and Tiele had both married and had children; that Kai had completed his tour in the army, Helga was engaged, and Tutti had graduated from high school and was preparing to become a teacher like the others. But there was also terribly sad news, that Klemens had suddenly died from a quickly spreading cancer, leaving behind a wife and two children.

Heidi ached to confide in her sister about what troubled her, that life at times could be lonely. It worried her that Opa was no longer in the good graces of the authorities, but she sought comfort in the fact that despite their circumstances, Oma remained strong, refusing to let the regime get the better of the family. Most of all, Heidi wished she could tell her sister that Hanna inspired her, and that she admired her for having escaped. But Heidi knew she could never risk writing these words in a letter. In light of the Stasi's ongoing campaign to slander Opa, Oma told Heidi to hold off on sending any letters to Hanna.

No one in his right mind dared poke fun at the regime. But now, at his Saturday-night card games, despite Oma's best efforts, Opa's frustrations came to a boiling point.

Knowing full well there were informants in his midst, one night he threw caution to the wind and ridiculed the local authorities, calling the regime a bunch of bumbling idiots. That comment silenced the room. After a long, deafening pause in which no one dared look at anyone else, Opa shrugged his shoulders and threw in his cards, motioning for the players to simply continue the game, which they did. Later in the game, half-jokingly, Opa called the mayor, one of his fellow card players, a cheater. But he really went overboard when he told a joke about East German leader Ulbricht, calling him a "backward stooge." Stunned at Opa's audacity, no one dared react. Opa, however, busted out in laughter. One of the other card players, his neighbor and longtime friend, realizing he had taken it too far, took hold of him by his collar and pulled him up saying, "That's right, old boy, I think you're done for the night. Let's get you home."

Before depositing him at home to Oma, his friend yanked Opa close, looked him dead in the eye, and told him that he was acting reckless and asking for serious trouble.

It didn't take long for the authorities to call Opa in. They told him they were fed up with his attitude.

Then one of them slid a piece of paper across the table to Opa. It was a letter he had written to Ulbricht in which he complained about the failings of the local Communist Party leadership.

"Is this your signature?"

Opa returned home with a blank stare. He had been dismissed from his job as headmaster and teacher, officially denounced, and kicked out of the Communist Party.

After several weeks at home, fuming, worried, and thinking things through, one afternoon he rose from his chair in his study and pulled himself together. Then he walked back to the village headquarters and told the authorities that they had made a mistake dropping him from the Party. Citing his long and dedicated service to his community, he asked that his membership and job be reinstated.

The authorities responded by telling him that somewhere along the line he had lost his loyalty, adding "your performance is no longer in the spirit of progress," and that Opa's continued presence in the village would affect the other villagers.

And that is how Oma and Opa, together with the only child left at home, Heidi, were unceremoniously banished to the tiny hamlet of Klein Apenburg, nearly seventy miles away and in the middle of nowhere.

13

ONLY PARTY MEMBERS SUCCEED

"WE HAVE EACH OTHER"

(1966–1969)

I do not weep: I loathe tears, for they are a sign of slavery.
— *German artist Max Beckmann*

*K*lein Apenburg was located in a remote region of the Altmark district of Saxony-Anhalt. Surrounded mostly by farmland or vacant stretches of undeveloped land, the entire community consisted of seven modest buildings clustered around one dusty cul-de-sac. Five of those buildings were dwellings, one an abandoned stone church with a bell tower, and the other an old, decrepit wooden barn. Everyone in the hamlet was elderly and a few infirm. One octogenarian had lived in the tiny settlement for decades, but the others had arrived over the years alone or in pairs. Now the tiny community of retired farmers, a soldier, and factory workers, some with spouses, some alone, added a former headmaster and schoolteacher's family to their little colony. Other than sixteen-year-old Heidi, there were no children.

The house in Klein Apenburg

The new neighbors came to welcome them, bringing flowers and potatoes from their gardens. But Opa was in no mood to be friendly and remained sitting on the front stoop of their tiny new house, not even bothering to acknowledge them. Oma thanked them, explaining that he wasn't feeling well. They nodded and seemed to understand.

The notion of marginalization from mainstream society was a blow to the whole family. Roland, Kai, and Manni came to rally around their father and help Oma and Opa move in. The grown sons tried their best to console their father but saw that he was utterly depleted.

Oma refused to cry about their circumstances. She told her boys that Opa needed time to recover and that, after a while, he would be all right.

"*Da müssen wir durch*. This is just something we have to go through," she said. "We will be fine. We are strong. We have each other. Nothing will break us, neither this nor anything else. This family has far too much to be proud of. We are far above all of this."

*T*he boys moved in Oma and Opa's belongings, including some furniture and Opa's books. The tiny dwelling could only accommodate about a quarter of what they had owned, so much of Opa's study, most of his library, his collections of pressed wildflowers and geologic specimens, and some family treasures like the Heidelberg Castle kit model, were divided up among the children. But framed family photos and albums made it into the inventory. Believing it would help to revive him, Oma had insisted Opa's Schimmel piano be brought to Klein Apenburg. The rest of the family chipped in to pay for a truck to haul it.

After his sons left, Opa felt lost. Suddenly stripped of his purpose in life, and now physically separated quite a distance from his grown children, he plodded through the next couple of months in a haze.

*O*ma took charge. She tended lovingly to Opa but wasted no time feeling sorry for their situation and instead set about organizing the new house and creating a new life. With no running water, she shrugged off what she saw as minor inconveniences and set to work. After surveying the topsoil on the plot of land on the side of the house, she began tilling the soil for the foundation of a new garden. She put up pictures of the family all around the house. Refusing to let Heidi dwell on the family's misfortune, Oma handed her a rake.

"A garden can always change things," she said smiling as she knelt down and dug her hands into the earth. "With new seeds there are new beginnings."

*B*y late autumn, Oma was treating Opa with a steady regimen of curative herbal teas, hearty vegetables, and fresh air. She insisted he go outdoors every day and so he sat on the wooden bench at the far corner of their lot. With his blank expression it was hard for Oma to know what was going through his head. She reminded him constantly about his many achievements and contributions to the community as a successful leader and beloved teacher who had selflessly educated and helped so many over the years. At night she put him to bed, gently stroking his head and reminding him, "We have raised a good family. You have been a wonderful father and teacher. We will be fine."

*M*y grandparents and Heidi learned to live simply in Klein Apenburg. As part of Opa's retirement, his income was curtailed and they now had to live on vastly diminished resources and a meager state pension, which amounted to only a few dollars a month. What Opa could once upon a time afford for an entire family of eleven, he could now barely afford just for the three of them. Living in the tiny, isolated community they had only the very basics that they could grow themselves or exchange with their neighbors. They lived simply and sparsely in a small, plain, rustic house with a crude outhouse and a rusty water pump outside the kitchen door.

Though Oma had never spent much time on a bicycle, now older and less agile than she once was, she fought off isolation by occasionally riding the old family bicycle to the town of Apenburg, a four-mile round-trip, to see what she could procure for the family.

The high school too was in Apenburg. Heidi made the journey using that bicycle every day that fall, but when winter set in, espe-

Oma, behind the Iron Curtain

Courtesy of the Willner family

The Heidelberg Castle, West Germany
Courtesy of Reinhard Wolf

The Hoheneck Castle, East Germany
Courtesy of Stiftung Sächsische Gedenkstätten

In East Germany, two years after Hanna has fled, a new family portrait is taken. (*Front row, from left:*) Klemens, Oma with baby Heidi, Opa with Helga, an aunt with Tutti, and Kai. (*Second row, standing:*) Roland's wife, Roland, Tiele, and Manni.

Courtesy of the Willner family

East Germans, young and old, flee into West Berlin on August 12, 1961.
Courtesy of Habans/Getty Images

East German soldiers and workers build the Berlin Wall on August 13, 1961.
Courtesy of Lackenbach/Getty Images

Two families escape
to the West in this
homemade balloon.
Courtesy of Günter Wetzel

An unsuccessful escape attempt, Invalidenstrasse, East Berlin
Courtesy of Polizeihistorische Sammlung

Beyond the Berlin Wall
Courtesy of Mathias Donderer

Operations, Tempelhof Airport, West Berlin
Courtesy of the author

Checkpoint Charlie

Courtesy of Roger Wollstadt

East German border guards look into the West.

Courtesy of Keystone-France/Getty Images

Oma in her garden in Klein Apenburg
Courtesy of the Willner family

Albert with Opa
Courtesy of Michael Nelson

Albert with the family in Klein Apenburg
Courtesy of Michael Nelson

Paradise Bungalow in 2005
Courtesy of the Willner family

Heidi in Paradise Bungalow in 1984
Courtesy of the Willner family

Cordula wearing the bathing suit
Courtesy of the Willner family

Nina
Courtesy of the author

Cordula at international
road race
Courtesy of the Willner family

U.S. president
Ronald Reagan (*right*)
with Soviet leader
Mikhail Gorbachev.
Courtesy of AFP/Getty Images

Anniversary parade, Karl-Marx-Allee, East Berlin in October 1989:
"40 Years East Germany"

Courtesy of Picture-Alliance/dpa

Soviet leader Gorbachev and East German leader Honecker attend
forty-year celebration of East German rule, East Berlin in October 1989.

Courtesy of Picture-Alliance/Sven Simon

Heidi and Reinhard leave the East in the Skoda.
Courtesy of the Willner family

(*Top right:*) Hanna reunites with Manni.
(*Middle and bottom right:*) Heidi (*on the left*) reunites with Hanna having only
met once before, briefly during the Cold War when Heidi was five years old.
Courtesy of the Willner family

Albert, Cordula, and Nina run the Berlin Marathon in 2013.

Courtesy of Marathon Foto

Family reunion in 2013

Courtesy of the Willner family

Hanna and Heidi in 2015

Courtesy of the Willner family

Courtesy of the Willner family

cially after a heavy snowfall, she had to go by foot, and sometimes showed up late to school even when she had started out early. She worked hard to get excellent grades so that Oma and Opa would have something to be happy about, and so that she would not add to their burden by doing poorly.

On the long, freezing-cold walks to and from school, she had plenty of time to reflect on her circumstances. Heidi's feelings about Opa ran the spectrum between deep sympathy for him and anger that he had caused their isolation by railing against the regime and losing his temper. She wondered if, outside East Germany, they banished people to remote places when they spoke up against their government.

Occasionally, on her secluded treks to school, Heidi's mind wandered to thoughts of her oldest sister. The more Heidi thought about Hanna, the more she came to admire her for having had the courage to run, even if it meant leaving everyone behind. Hanna had been a risk taker, unafraid in the face of danger, and had done what so many others inside East Germany wished they had the courage to do.

And so during that first winter in exile, somewhere on the long road between Klein Apenburg and Apenburg, Heidi came to virtually idolize Hanna for her strength and daring. Having physically emulated her sister throughout her childhood, as a teenager Heidi now resolved to pattern her character after Hanna's. She tried to embody what Hanna would think and feel in every situation, how she would react, what she would or would not say. Imagining her sister would be proud of her, she vowed to carry herself with dignity no matter what might come.

With a renewed sense of direction, rooted in her sister's courage in having mounted the ultimate act of disrespect and defiance against the regime by escaping, Heidi became emboldened. She catapulted Hanna's image to new heights and by the end of that desperately lonely year had come to think of Hanna as nothing less than a legend.

*A*s winter drew on and snowfalls became heavier, it became too difficult for Heidi to get to school, and since Oma and Opa had no money to board her in Apenburg, she had to drop out of school altogether. She spent the long cold days at home, working around the house, helping Oma and Opa, cooking, cleaning, carrying in the wood and coal, playing cards with them, and having long talks with Oma about life. They did a lot of needlework that winter, crocheting blankets and embroidering the edges of handkerchiefs. When spring came around, having missed too much schoolwork to reintegrate back into her class, Heidi remained at home, splitting firewood, helping Oma work the garden, tending to Opa, and wondering about her future.

For a time, Opa remained morose, often wearing the gaze of a man who had emotionally departed. But after nearly a year in Klein Apenburg, thanks to Oma's care he became more accepting of their circumstances. He spent his days reading and helping Oma in the garden; from time to time he even chatted with the neighbors. After a while Opa even joined them in their card games. Though he seemed to be adjusting, when Oma tried to get him to play his Schimmel, after several halfhearted attempts he said it didn't sound right, that it was out of tune.

Turmoil spread around the globe. In Asia, American involvement in Vietnam peaked when the Vietcong launched the Tet Offensive, handing the United States a political defeat. Communist China detonated its third nuclear bomb. U.S. forces invaded the Dominican

Republic to prevent a communist takeover like the one that had taken place in nearby Cuba.

In Czechoslovakia in 1968, Soviet, East German, and other Warsaw Pact troops crushed the "Prague Spring." Like the East German uprising of 1953 and the Hungarian revolt of 1956, the Czechoslovak uprising was another lesson to the people of Eastern Europe not to step out of line or challenge communist rule. And so the leaders of East Germany got the green light to continue unhindered on their path of oppression.

The race for superiority in space continued as both sides spent huge resources to achieve new heights, peaking anew in 1969, when the United States landed a man on the moon.

I was eight years old when I watched in awe along with my parents and 500 million others around the world as Neil Armstrong took the first steps on the surface of the moon. It was a tribute to perseverance and innovation and a victory not only for Western technological advancement in space, but also for freedom and democracy.

By the late 1960s, the East German leadership realized that they had to do something to improve their country's reputation with the outside world. Their plan would have to be something so enormous in scope that it would command international respect and put East Germany on par with other leading countries of the world. Thus a massive campaign was launched to dominate the world of sports, including the Olympic Games. East Germany's goal: to surpass West Germany in every form of competition and compete on the same level as the two world superpowers, each with more than 230 million citizens compared to East Germany's 17 million. It wouldn't be long, the regime believed, before everyone would sit up and take notice and East Germany, once considered a pariah, would earn the world's respect as a leader in international sport, and in the process, gain a platform to promote the achievements of East German communist society.

They launched their campaign with great discipline and fervor.

Propaganda posters and banners suddenly sprang up, compelling children everywhere to outrun, outrace, outswim, and out-throw everyone else. Every city and town, every gymnasium and school, hailed the new future of competitive athletics.

The leadership formed a sports ministry and poured money into athletic programs throughout the country. Suddenly, in schools, at FDJ and Young Pioneer meetings and at summer camps, sports took center stage. Prompted by new standards outlined by a national sports board, gym teachers and coaches set lofty goals for their students that stressed fitness, strength, and stamina. Athletic clubs across the country received lavish subsidies. The regime recruited the best sports trainers and talent scouts in the country and dispatched them to schools and gymnasiums to find and recruit the best young athletes East Germany had to offer: the fittest, the strongest, and the fastest, for swimming, gymnastics, weight lifting, ice-skating, and cycling.

By spring, though Opa had lost weight and had clearly aged, he seemed to finally be adjusting to his new circumstances. With Opa somewhat stabilized, Oma turned to Heidi, now eighteen years old, and told her that she needed to leave Klein Apenburg and prepare herself for life on her own.

Oma and Opa could not afford teachers' college for Heidi and, given Opa's outcast status and Heidi's failure to finish high school, she would not have been accepted for training. So instead, after two years in Klein Apenburg, she went north to Salzwedel to train as a stenographer. After several months, she completed the program and, along with her credentials, was given a chance to join the Communist Party. She quietly discarded the application and set off to find a job.

Prospective employers always seemed to ask the same two questions, even before inquiring about her skills and qualifications.

"Are you a member of the Communist Party? Do you have any relatives living in the West?"

Heidi answered truthfully each time. She was not a member of the Communist Party and she did have relatives in the West. When asked if anyone in her family had fled, she answered in the affirmative. Everywhere she applied, she was rejected.

Heidi returned to Klein Apenburg. Opa was livid. He berated Heidi, telling her that she had no sense of reality.

"Without joining the Party," he bellowed, "you have no chance at a decent life!"

Always good natured and never rebellious, she mustered the courage to challenge him for the first time in her life. "Were they *so* good to you?" she asked. "You gave them the best years of your life and look where you are!"

Opa hollered back, "You're too young to understand! Why are you *so stubborn?*"

Oma put her hand on Opa's shoulder, patting him in an effort to get him to calm down. He took in a deep breath, then looked into Heidi's eyes.

"Do the right thing, Heidi," he said, struggling to keep control of himself. "Only Party members succeed." And with complete conviction, he added, "Sign or you will achieve nothing in your life, I promise you that!"

Heidi continued to protest: "What does it matter, Papa? Everyone in the Party still has to look over their shoulder and watch what they say and do. What does it matter?"

"It matters! It matters!" Opa yelled. Raging now, he shouted, "People succeed because they play by the rules! Can't you understand that?"

Now shaking with anger, his wild expression boring through her, he said, "You are a troublemaker, just like Hanna!"

*O*ma and Heidi spent that evening lingering on the stoop outside, which looked out over the dirt path entrance to the hamlet and on to the old, decaying church across the way.

"You'll be all right," Oma said, her comforting words belying her concern that Heidi's decision not to join the Party would indeed affect her last child's future.

"You always told me to do the right thing, to be true to myself," Heidi said, looking at Oma for guidance.

"Do what you think is right," Oma reassured her. "You will be fine."

An idealist by nature, Heidi, determined to prove that the caliber of character she had and quality of work she could produce were far more important than mere membership in the Communist Party, returned to Salzwedel and set off once again to find work. For months she looked for a job, hoping to find some office that needed a stenographer, but time and time again she was rejected.

During those months, Heidi began to understand what Opa had tried to tell her. Lack of Party membership would limit her opportunities. At some point, feeling distressed and wanting to make contact with Hanna, at the risk of drawing even more negative attention to herself, she braved a short letter. Like Heidi, everyone in the family figured that they might have a better chance of a letter making it to Hanna if they kept things vague and even flattered the regime.

"My dearest Hanna," Heidi wrote, "We are well and happy. Our parents have moved to a smaller place. Life for us here is good. Here is a picture by which you can remember me. Please do not forget me." That letter and photograph made it out.

In the United States, I turned nine years old. My father, Eddie, retired from the army, ending his career as a major, and went on to work as a civil servant for the U.S. government. My mother, Hanna, was a housewife and part-time German language teacher.

By now there were five children in our family, born one after the other, the product to a large degree of my parents' intent on replacing the families they had both lost. We were a close-knit family. Albert was the oldest, then me, Marcel, Maggy, and Sachi. At home, my parents spoke German to each other. There was no expectation to learn German but if we wanted to know what our parents were talking about, we worked harder to pick it up. We had a happy American childhood, attended public schools, took piano, violin, and art lessons, played sports in school, went to Jewish Sunday school, and spent every day during the summer months at the local swimming pool. Like other American families, we took road trips to other cities, and occasionally went camping and canoeing on the weekends.

Oblivious of the strife engaging the world, we knew very little about our relatives in East Germany, and knew nothing about the repression they endured. Only occasionally, when I passed by my parents' bedroom, would I glance at the photograph of my aunt Heidi and the framed photograph of the whole family together marking

Oma and Opa's silver wedding anniversary, a picture in which only my mother was absent. And I would wonder how all my relatives were coping behind what I had by now learned was the Iron Curtain.

*A*fter several months, the tide suddenly turned for Heidi. At the age of nineteen, she met a twenty-three-year-old conscripted NVA soldier who was finishing his mandatory tour in the East German Army with duty at a casern near the East–West Inner German border, as an electrical technician.

They met at the local dance hall, Der Schwarzer Adler, the Black Eagle. Dressed modestly in a beige blouse and black skirt, she sat quietly at a table with a few classmates from her stenography class. After a while, an affable, sandy-blond-haired soldier in uniform came and asked her to dance, and she accepted. He could dance well, she couldn't, but they found a way, albeit a little clumsily on her part. After the song ended, Reinhard introduced himself. They talked a bit, then she thanked him and rejoined her friends. Though he was immediately smitten, they parted ways without any further plans to meet up, but by the next weekend they were both at the dance hall, happy to see each other again. Unbeknownst to Heidi, as no buses ran from his garrison to Salzwedel, Reinhard had made the hour-long trip by foot hoping to see Heidi once more. After that, they met up at the dance hall regularly, by then Reinhard having had his parents send the family bicycle so he could make the trip more quickly.

Reinhard was a well-intentioned, honest soul, with an optimistic outlook and a quiet, steely-eyed determination. Like Heidi, he was not a member of the Communist Party and never intended to be one. He thought she was his ideal match; she had a lightness of spirit

yet an understated resolve to live life on her own terms. He was en-
chanted with her zest for life and her independent streak.

Heidi found in her new love a kindred spirit. After a short court-
ship, the two married. With at least a two-year wait for non–Party
members to get an apartment, they moved in with Reinhard's parents, who lived in Stollberg, the picturesque valley that lay at the base
of the Hoheneck Castle Women's Prison.

In nearby Karl Marx City, Heidi set out once again to find work.
After several more months of searching, she finally found a job as a
secretary with a state auto part design and engineering firm. To her
surprise, without even asking if she had family members in the West
or if she was a Party member, the office leader, Herr Meier, hired her.

Grateful to Meier for giving her a chance, Heidi threw herself into
her job and became a model worker. Within a month, she became in-
dispensable to her boss, who quickly learned to rely on her consistent
productivity and strong work ethic.

After serving his tour in the army, Reinhard, who had excelled in
mathematics in school and had hoped to pursue higher education
as an engineer, due to his lack of Party alliance, instead became a
field electrician. Accepting their modest positions in life, Heidi and
Reinhard found their way; an inner strength between them began to
shape their lives.

Two years after they married, the two were excited to be assigned
a small flat in Karl Marx City in one of the thousands of brand-new
Soviet-style, prefab concrete-block high-rise apartment buildings that
were sprouting up throughout East Germany. While Party leaders received larger quarters and didn't have to wait as long, theirs was a standard apartment in the industrial sector of the city. Heidi and Reinhard
could look across the street into the apartment building housing Soviet

officers. Their Soviet neighbors kept to themselves and never mingled with the East Germans who lived only a few yards away.

After several years living on a small income, Heidi and Reinhard had finally amassed enough money to order a car. With few East German cars to choose from, it was either the two-stroke Wartburg or the Trabant, or Trabbi, known as the Cardboard Car. Neither car was fast, and both puffed out loads of pollution, but at about the equivalent of $2,500, the Trabbi was far less expensive than other East European brands. Instead of choosing an East German model, Reinhard excitedly put in his application and a down payment on a Czechoslovakian-made Skoda, a better, more sturdy vehicle, but also more expensive and harder to get. For non–Party members there was a fifteen-year wait.

That same year, my parents got a new Ford station wagon and bought a large house on a quiet cul-de-sac in a tranquil suburb of Washington, D.C. We were an active American family with a household full of spirited, energetic children and two large German shepherds when my mother, Hanna, learned that she was pregnant with her sixth child.

At around the same time, in Karl Marx City, Heidi learned that she was pregnant with her first.

14

A MESSAGE WITH NO WORDS
OMA'S LOVE FROM AFAR
(1970–1974)

Youth fades; love droops; the leaves of friendship fall. A
mother's secret hope outlives them all.
—*Dr. Oliver Wendell Holmes Sr.*

*I*n the winter of 1970, in a state maternity hospital in Karl Marx
City, Heidi gave birth to a healthy little girl. She wrote a letter to
Hanna that made it out. I was thrilled to hear that I had a new
cousin, even though I understood by now that we would likely never
meet. Her name was Cordula, which means Little Heart. At nine
years old, I couldn't always remember the name Cordula, but I could
remember Little Heart, so for a while I called her that.

Over the next weeks, my mother, Hanna, knitted a cotton blanket, a tiny cardigan, baby booties, and a little bonnet. I watched as
she folded the little garments and placed them carefully into a box.
She too had recently given birth, and inside the package she placed
a card with a handful of pictures, including one of her six children.

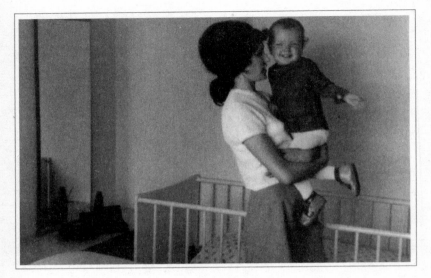

Heidi with baby Cordula, Karl Marx City, East Germany

A few months later, we were surprised to receive a letter from Heidi saying everything was fine, but it made no mention of the contents of the package except to thank Hanna for the baby bonnet and the photographs.

Because they did not want to call undue attention to themselves, the letters the family sent out were purposely subdued and vague, trying not to give censors anything to object to or the Stasi anything to exploit. In reality, however, the family in the East was thrilled to get the photos of Hanna and Eddie and their six children.

The whole family viewed it with curiosity and great amusement. Oma carefully examined the image of our six happy little faces. Heidi studied the picture for a long time, poring over the faces of her nieces and nephews. But she became melancholy when she realized that she and her own children would likely never know Hanna's children.

Hanna's six children in suburban Washington, D.C.

Heidi longed to share the photograph with her friends, neighbors, and coworkers, but she knew better. The mere act of showing the image of her American sister or her sister's children to anyone would be a mistake and could trigger unwanted attention from the authorities. No one knew who the citizen informants were, but everyone knew it was better to be safe than sorry.

By the early 1970s in Klein Apenburg, Oma had nursed Opa back to health, and he had made great strides in overcoming his feelings of loss over being let go as headmaster. Time, the onset of older age, and Oma's gentle touch had softened him. He was happiest when any of his now fifteen grandchildren came to visit, and he was especially smitten with the latest addition to the family.

Cordula was a sweet, happy Gerber baby look-alike. She grew into

Hanna knit a blanket for Heidi, then sent it, but Heidi never received it.

a cherubic, apple-faced toddler with curly, honey-golden locks. Opa was captivated by her little girl charms and her vitality and completely doted on her when she came to visit. Taking her little hand, he would show her what was growing in Oma's garden, asking her to help him pick berries or gather up twigs from the yard. He hoisted her up so that she could pluck plums and apples from the trees. At night he would bring her outside and show her the stars, taking his time to tell her the stories of the constellations. He taught her the names of flowers and trees, insects and butterflies, and was always patient in trying to understand and answer any questions she had, the sound of her little voice like music to his ears. She in turn gifted him with pinecones she had gathered, and with stones and pebbles she had found in the yard. Their relationship was a sweet one, nurturing them both and bringing much-needed joy to Opa's otherwise diminished life.

Occasionally, several of Oma and Opa's grown children and their families came to visit at the same time, which especially delighted my grandparents. Because of Klein Apenburg's remote location, it took almost everyone a full day or two of travel.

They descended on Klein Apenburg in threes and fours, spreading out in every corner of the house, claiming space and setting up wherever they could find room, arranging makeshift beds on couches or on the floor. While the men chopped wood and stacked it against the house, teasing and joking as they did so, the women and children worked in the garden or helped Oma seal jars of gooseberries and currants as they gabbed about their lives in the cities and towns where they lived. Outside, the children played in the garden or chased one another around the yard; indoors, they tinkered on the Schimmel, snooped through their grandparents' personal things, or perched in the kitchen waiting for cakes to come out of the oven.

The little ones loved to take long walks with Oma and Opa, all of them always clamoring and pushing to be the one to hold their grandparents' hands. The girls would pick flowers along the way, which they would then pool to make Oma a daisy-clover "fairy wreath" crown, amused to no end when she wore it well into the evening hours. All of the grandchildren liked to help Oma in the garden, which had grown into a virtual botanical wonderland, a lush panoply of flowers, fruits, and greens where she grew foot-long carrots, white cabbage, potatoes, sweet strawberries, rosy red currants, and plump, juicy gooseberries. Cordula especially liked to man the water pump for the cooking pot and the watering can. She reveled in discovering new fruit and nut trees that grew in Oma's plot, and especially loved harvesting the peas from the garden, where she often ate the fresh, raw peas right out of

On "Opa's resting place" bench with baby Cordula

the pod, whimsically sweet childish moments that never failed to delight Opa.

Opa spent a lot of time sitting on his bench on the edge of the garden, smoking his pipe, and before long the grandchildren began calling it "Opa's resting place." Just sitting next to their grandfather became a coveted treat, a place of honor, and they sometimes tussled with one another, competing to be the one to sit closest to him. Sitting on that bench next to her grandfather would become one of the highlights of Cordula's visits to Klein Apenburg, among her most cherished memories with Opa and likely some of the most treasured times of his life as well while in exile.

At the age of sixty-seven, Oma was diagnosed with diabetes. The doctor gave her some medicine and recommended she eat grapefruits, which were impossible to find. Heidi wrote to Hanna of Oma's diagnosis and the recommended treatment. In the United States, Hanna boxed up a case of Florida grapefruits and sent them to Oma, but that package, like so many others, never arrived.

*I*n late 1972, the outpost community of Klein Apenburg finally got a telephone. An old, black antique Bakelite ball-bearing dial telephone was installed in Oma and Opa's house. Meant to be shared by all the hamlet inhabitants, it had a very loud ring that roused attention throughout the whole village every time it sounded.

Tiele, a kindergarten teacher with a clean record, seemed to have the best success getting letters in and out to Hanna. One day my mother received an innocuous letter from Tiele in which she wrote that she was pleased that she could now reach Oma and Opa by telephone. On the inside flap of the envelope she had written some numbers, which Hanna initially discounted as scribble, but after a while came to believe might be her parents' telephone number.

In the summer of 1973, fifteen years after she had last seen or talked to Oma, Hanna got the international operator on the phone and gave her the number. She knew full well that this one phone call could jeopardize all future contact or otherwise put the family in danger, but at that moment her desire to speak to her mother was overwhelming. She tried to calm her nerves and pulled her racing thoughts together.

I was twelve at the time. Barely able to contain my excitement, I stood nearby, bouncing up and down on the balls of my feet. My father, Eddie, and my five brothers and sisters huddled around our mother. We all waited nervously. The rings seemed to last forever.

A German operator picked up and the international operator

tried to put the call through, but she was unable to make a connection. My mother second-guessed why the call would not go through, was quick to attribute the problem to Stasi interference, and began to reconsider. She wavered, wondering if calling was the right thing to do or if the phone call was too risky for the family in the East.

But Eddie said, "Try again."

Again the call did not go through. Then she became determined to make contact. For the next hour, she doggedly tried over and over and over again. Suddenly, amid clicks and buzzes, a connection. I was beside myself with glee, hanging on my mother's every breath as the phone on the other end started to ring. With every ring came a jolt of anticipation.

Finally, on the other end, someone picked up. We all looked at one another. My mother was stunned silent. Then almost inaudibly, in German, she uttered, "Mutti? Mutti, it's me."

"Who?" Oma called back in German, through the static interference. I was very excited to hear Oma's voice for the first time.

"Who is there?"

"It's me, Hanna," she said, almost whispering now.

"Who?" asked Oma loudly through the crackling. "I can't hear you. Who is this? Speak up. Where are you calling from?"

But Hanna couldn't speak. Then she found her strength and let out a breath: "Mutti, it's Hanna!"

No answer came from the other side. Time stood still. We hung on, waiting, waiting for more.

"Hanna?" Oma finally said as if in a daze. "Hanna?" she said again softly, her voice quivering.

My mother was too overcome to speak further. They stayed on the phone in silence for what seemed like several minutes. Then, through

Oma, Heidi, Opa, and Cordula in Klein Apenburg

her crying, Oma finally said, "We miss you so much," to which my mother could say nothing more. Soon after that, the phone call was interrupted by high-pitched beeping noises and the line went dead. My mother pulled herself together and dialed the operator again, and again and again, but they were unable to get through a second time.

She had always been so far away from us, but for a fleeting moment, Oma had been right there. My Oma, the woman with the tranquil smile sitting askew in the polka-dotted armchair, had been right there with me in our kitchen.

With East Germany's national sports program cranked up and now well under way, athletes started to show remarkable results at international competitions. In just a few short years, they were already competing alongside world-class competitors who noticed that, for such a small country, East Germany had a great deal of emerging talent. The leadership was pleased with the determination of the trainers and the progress of the athletes in the program. Encouraged by their successes, scouts, coaches, and gym teachers fanned out across the country, scanning schools and sports halls to find the most athletically talented six- to ten-year-olds and whisk them into the system.

By now, East Germany had a new leader. Erich Honecker, the man who had successfully launched the communist youth movement and served as oversight authority on the construction of the Berlin Wall, took over as general secretary. His portrait was put up in every office, school, factory, and border watchtower around the country.

Chosen in part because of his uncompromising loyalty to the Soviet Union, Honecker was seen by Moscow as a reliable man who could get things done. He entered office ready to make major changes and intended to make full use of the secret police to reach his goals. Confident and composed, he looked like a modern European leader but would quickly emerge the epitome of an iron-fisted totalitarian dictator.

In one of his first decrees as new leader, in part to win over his people's loyalty, Honecker introduced what he called "consumer socialism," promising to raise the standard of living and satisfy the population's desire for more consumer products. Where store shelves were empty before, by the early 1970s more goods became available to the general population. While there was no marketing, and East German brands were not high quality and had no fancy packaging—the regime favoring muted tan and pale green hues—the products

Official portrait of East German leader Erich Honecker

gave the people something where there once was nothing.

Refrigerators and washing machines, cosmetics and kitchen appliances all became available, as did clothing and more food options, including chocolates, jarred marmalades, and pickles, noodles, even "champagne," a mild, bubbly wine. Coca-Cola was not available but the herb-flavored Vita-cola was. More clothing became available though it sometimes fell apart after one washing. While Japanese Sony radios were not to be found, Stern-Hobby, the no-frills East German brand, could now be purchased. Heiko pens, while not as fluid or long-lasting as Bics, worked fine for a short time.

Heidi and Reinhard saved their money and ordered a Scharfenstein refrigerator, easy to choose since it was the only East German model available. Though it was not much more than a sheet metal box made from cheap materials, and hardly worked properly, never-

theless they now had a refrigerator. That purchase was followed by a
Schwarzenberg washing machine, which operated only half the time.

Some excellent, finer-quality items were also produced. The best
products, beautifully crafted Diamant bicycles, Plauen lace curtains,
Glashütte clocks, Zeiss binoculars, or Lauscha glass, were too expensive for the average East German and made almost exclusively
for export to foreign countries. Even though Christmas was not officially sanctioned as a religious holiday, Christmas decorations, such
as wooden nutcrackers, angels, intricately carved Erzgebirge *schwib-
bogen* candelabras and pyramids, and nativity scenes, as well as *Stol-
len*, a traditional Christmas cake, were produced in large part for the
foreign market for its foreign currency.

Even as more products became available, people often still stood
in long lines in state shops as part of their daily routine to buy staples
like butter, eggs, and fresh vegetables, which often ran out by the
time they reached the head of the line. Though potatoes and apples
were available, bananas, citrus fruits, and vegetables like broccoli
were hard if not altogether impossible to find. Some produce came in
from other communist countries—oranges from Cuba, for example,
were a rare treat.

Though marketing was nonexistent, quality was lower, modern
design innovation was absent, and selections were limited, nevertheless, as new products hit the shelves, many praised their new leader
for making improvements. Just as Honecker had anticipated, the
new products provided East Germans some measure of hope that
things were looking up.

In Klein Apenburg, Opa wrote to Hanna about the sudden availability of goods. Not surprisingly, the authorities let that letter make

it out of the country. In his letter, Opa wrote that the family had gathered for Christmas to celebrate and exchange presents. Describing the gift exchange, he wrote:

> *The girls received hand towels and handkerchiefs, all in beautiful designs and colors, also underwear sets, cologne, and boxes with soaps. Your mother received a colorful apron and coffee from Tutti, and from Tiele a slip and coffee. We also made a bunter Teller [a Christmas plate] this year, with chocolates, cookies, macaroons, walnuts, and hazelnuts. It is all very exciting. We have never had this many presents before.*

As "excited" as he sounded, at the end of his letter, he added a cheeky message:

> *Your mother just bought some new curtains. The other ones dissolved in the laundry and looked like a noodle soup. We are hopeful that the new ones are of better quality.*

My mother, Hanna, was overjoyed to get this news and to learn that the family seemed to be faring well.

Heidi and Reinhard avoided joining the Communist Party and tried to keep a low profile. They worked hard at their jobs but neither saw the promotions or advancements that their Party member colleagues received. Nevertheless, they stayed firm to their commitment to their ideals, in their quiet way of protest of the system that they did not believe in. They would work hard, do their best to be good to neighbors and coworkers, but not toe the communist line. They understood the consequences and learned to live within their means; they never complained, came to appreciate what they had, especially the strength of family, and were determined to make their lives meaningful on their own terms and within their circumstances. And, with another child now well on the way, they were determined to make the best that life could give them.

"We have a peaceful life," Heidi wrote to Hanna, and meant it. "We enjoy every day."

*I*n the early 1970s, the United States and the Soviet Union entered a new era in which for the first time the two superpowers seemed open to working together to ease tensions. They began talks to reduce nuclear arms inventories; détente aimed to thaw relations. President Richard Nixon became the first U.S. president to visit the Soviet Union and Communist China.

*I*n their apartment in Karl Marx City, as a part of their evening routine, Heidi turned on their flickering black-and-white television, allowing little Cordula to fall asleep to the soothing sounds and spritely antics of a tiny wooden, stop-motion-animation puppet. With sweet children's voices singing tender songs of peace and contentment, Sandmännchen (Sandman), with his white yarn hair and beard, pointy hat, and curled shoes, took the little viewers around the world, to places they could not go themselves, places they could only dream of, riding on a cloud, piloting a rocket ship or helicopter or riding a magic carpet to meet a princess on the Golden Horn, nomads on the Kazakh steppes, reindeer in boreal Siberia. When he grew tired, Sandman sprinkled glittery golden magic sleepy dust

Sandman is free to travel the world and fly to other lands.

from his sack, encouraging children to set aside their worries and drift off to a carefree slumber.

> Children, dear children, that was fun.
> Now, quick to bed and sleep tight.
> Then I will also go and rest. I wish you a good night.

As the population lived sparsely, under the watchful eye of the secret police and a rigid code of silence, East Germany's political elite was living it up in luxury.

In a heavily guarded, secluded forested area spanning a square mile, privileged members of the regime lived on an exclusive estate that bordered a pristine lake. Unbeknownst to the people of East Germany, who assumed their leaders lived similarly spartan lives, the Wandlitz Forest Settlement housed some twenty mini-mansions complete with wide, manicured lawns and gardens. The compound was secured by a concrete wall with a sophisticated alarm system and guarded by more than a hundred heavily armed troops from the elite Feliks Dzerzhinsky security force. Instead of Soviet Chaikas, there were expensive luxury Volvos in the driveways.

Inside the houses, the communist leadership enjoyed spacious rooms decorated with marble from Italy and Renaissance furniture from France, and used appliances from West Germany. Foreign goods and foods filled the compound's shops. Honecker, Stasi chief Mielke, and others enjoyed access to a state-of-the-art medical clinic, a cinema, swimming pool, saunas and spas, a recreational shooting range, a sports field, tennis courts, a restaurant, and even an underground bunker built to protect them should war break out. While millions of East Germans stood in lines for lesser-quality food, cooks on the compound prepared exquisite gourmet meals made from imported delicacies, accented by the finest French wines.

When Oma and Opa turned seventy, the local authorities told them that they were free to go. As pensioners, they could leave the East and travel or even emigrate to the West. They barely discussed it, not daring to take even a single trip out, fearing they would not be allowed to return, to ever see their loved ones again. And my mother did not dare to travel to the East to see them for fear they would somehow prevent her from leaving the country.

$\mathcal{R}$oland, by now well into middle age, had built a highly successful career as an educator in the communist system. Years earlier he had been promoted to director of his city school and he now had his sights set on the position of school superintendent. In the eyes of the regime, he had a great record and, by all accounts, his prospects for promotion were excellent.

The rest of the family made their way in East German society: Manni, Tiele, and Tutti as teachers at various elementary and high schools, and Helga as a day-care worker. All raised fine families with one or two children. Heidi and Reinhard worked at their jobs in Karl Marx City and had another baby. At the age of four, little Cordula became a big sister when Heidi had another girl, Mari.

After leaving the army in the mid-1960s, Kai and his wife had had four children. He and his siblings remained close, but after a while they noticed that Kai became more distant, and no longer reached out to his family. Concerned that no one had heard from him in months, Manni paid him a visit and found him in bed suffering from a rare blood disorder, which was believed to have been caused from his work at the Peenemünde rocket production facility in northern East Germany, where he had been exposed to dangerous chemicals. At the age of thirty-four, their little brother, Kai, was dying.

15

DISSIDENTS AND TROUBLEMAKERS
OPA COMMITTED
(1975–1977)

You do not become a "dissident" just because you decide
one day to take up this most unusual career. It begins as
an attempt to do your work well, and ends with being
branded an enemy of society.
—*Vaclav Havel*

*I*n 1975, the Vietnam War ended with the communist North
defeating the U.S.-backed democratic South; in Chile, a U.S.-
supported military coup overthrew a leftist president; in Ethiopia, a
Marxist junta overthrew a pro-Western monarch; and in Cambodia,
a communist regime allied with North Vietnam took power.

Détente continued. In an effort to end the space race, the two superpowers even agreed to pool their resources and scientific and engineering knowledge, and collaborated on the Apollo-Soyuz Project,
a joint U.S.-USSR space flight. With the adoption of the Helsinki

Accords in 1975, in which the Soviets promised to allow free elections in Eastern Europe, it appeared that warmer relations were on the horizon and things were heading in the right direction.

Following Moscow's lead, Honecker too sought improved relations with the West. The two German states signed a treaty committed to developing normal relations, recognizing each other's independence and sovereignty, and establishing a working diplomatic relationship. Relations marginally improved, if only on the diplomatic front. East Germany's true intentions for improving relations were less than entirely genuine, and they took advantage of improved ties by recruiting scores of spies within the West German government. In Bonn, Günter Guillaume, a close aide to West German chancellor Willy Brandt, was discovered to be an East German spy.

$\mathcal{M}$eanwhile, as Honecker's consumer socialism supplied the people with more goods and basic necessities, people were encouraged when the economy appeared to inch forward. Fed by borrowed money from the West, East Germany began to make improvements in its basic infrastructure, upgrading roads and building more new homes. Basic food stocks became more affordable, and there were upgrades in the social infrastructure as well: universal health care, maternity benefits, preschool programs. The regime trumpeted its successes to the world, highlighting the advantages of the socialist system. To his people, Honecker described East Germany as a stable, safe country with low crime rates, where people's needs were met, compared to West Germany's unhealthy hyperconsumerism, overconsumption, crime, and social unrest.

With things seemingly improving inside the East and little access to the realities of the outside world, for the first time many people came to believe that the regime now actually had a plan to make things better. By now some even considered themselves lucky to be part of a society they saw as orderly and peaceful, with leaders who were making progress and taking care of them.

In Washington, Hanna readied yet another package to send to her family, this time including a box of cigars for her father.

Opa wrote, "Thank you for the cigars, but since we could not afford to pay the customs, we had to return them." Not surprisingly, the cigars never made it back to the States.

"Maybe if you send ten cigars," Opa had written as a postscript, "we could afford to pay the customs." She did so but he never got the cigars.

But then, to the astonishment of families separated by the Wall, as part of the plan to improve East Germany's reputation, the regime selectively relaxed some of its laws about contact between citizens of West and East Germany. Suddenly letters and packages from West Germany were cleared to reach East Germans. My mother sent packages to a friend in West Germany, asking him to forward them on from his address in Frankfurt.

For the first time since the fur coat made it through in 1958, an unopened, intact box made it through to the family in the East. Oma sent a letter back to Hanna thanking her for the presents and detailing the exact contents of the package to let her know she had received everything: "sweaters and scarves, coffee, chocolate, grape-fruits and oranges, soap, lipstick, pantyhose, children's clothing," and—Opa had counted—"forty-eight cigarettes." At the end of the letter, clearly delighted, Opa had written, "I have already smoked one of the cigarettes."

For the first time in twenty-seven years, the gray fog of distance seemed to dissipate a bit and there seemed to be a real connection with the family.

My mother sent more packages to the East and, to her great joy, more letters came out. Over the next year, she received more letters and photographs than she had received in the nearly thirty previous years altogether.

"We received your package," wrote Opa. "It was undamaged and customs free. Thank you for the tobacco and especially for the cigars." Oma wrote, "Thank you for the cosmetics and creams, the statue of Buddha, the calendar, oriental golden wall-hanging and the knitting wool," adding, "I plan to make a dress with the beautiful flowered fabric." Hanna read and reread those letters. Was it possible, she wondered, that things in East Germany were turning? For the first time, she had regular correspondence with real news about the family. She was beside herself to get a tiny glimpse into their lives when Opa wrote,

At this time we are busy with canning green beans and cucumbers and are looking forward to a visit from Heidi, Reinhard and the girls. Despite the lack of rain, all the vegetables in our garden are still all right, so that our garden can still offer a lot of vitamins. . . .

Roland is a director and supervises seventeen teachers. All the other children are doing well. I am very proud of all our children. They are all very close, they work hard and aim high. Unfortunately they cannot correspond with you for official reasons, I'm sure you understand.

*H*eidi too wrote plenty of letters, in which she spoke of little Cordula, baby Mari, and Reinhard, and said that her job and life were

Letters and photographs from the family in the East in the mid-1970s

good. While Oma and Opa wrote most of the letters, a few of my mother's siblings wrote at least one letter, but everyone knew it was still too risky for working-age people to think their correspondence with their sister in America would not come back to haunt them. Hanna longed to hear something, especially from Roland, whom she missed terribly, but he was doing well within the system and had to be careful about making contact even when it appeared the regime did not seem to mind.

Everyone self-censored their correspondence—everyone, that is, but twenty-five-year-old Helga, who wrote:

I am proud of my country that I can speak freely. It is very difficult for me that I can only dream of seeing you again. This will always trouble me. I am not surprised when I hear that young people risk their lives to flee to the West. But that takes a lot of courage and I could never do it.

Surprisingly that letter made it out, owing possibly to less monitoring, since she worked in a fairly benign job as a day-care worker. Another letter came shortly thereafter, in which she wrote:

I no longer think . . . I can speak freely. I have a colleague who is very outspoken, also the hardest worker, an outstanding professional. He has said openly that he believes some things in the West are good. Now suddenly he cannot apply for a promotion. Now everyone talks negatively about the colleague. I have decided that one should look reality in the eye or, at least, not close the eyes to reality.

That letter was followed not long after by one from Opa, who wrote: "If you have not heard from Helga lately, it's because she has had some arguments with the mayor."

$\mathcal{D}$espite what appeared to be a softening of diplomatic relations between East and West Germany, the regime upgraded security at the Wall, creating the most sophisticated version yet.

Metal spikes were added, as were nail beds, and fences with touch-sensitive, self-firing mechanisms that set off alarms and fired rounds when they detected a presence along the wire. Heavy concrete watchtowers replaced the old wooden ones and loomed ominously over the death strip, giving border guards better views and easier lines of sight to their targets. Tripwires let off signal flares that alerted guards to the movement of an escapee; the sandy death strip was now wider and had floodlights that could highlight fresh footprints. Three-ton concrete barriers reinforced the concrete wall; electrified fences were added, and more attack dogs. Any attempt to cross the heavily fortified border was now suicidal.

By the mid-1970s, untold numbers had been shot dead at the Wall, including Johannes Lange, who was fired on 148 times by eight border guards before being hit 5 times in the head and body. The guards involved in that murder were rewarded with promotions and the Medal for Exemplary Service at the Border and were presented souvenir wristwatches.

With new fortifications in place, would-be escapees traveled to other Eastern Bloc countries, then tried to sneak through on what they thought might be an easier route, through Hungary or Bulgaria. Others came up with ingenious schemes. The Bethke brothers, Ingo and Holger, had escaped from East Germany but wanted to return to rescue their other brother, Egbert. Both flew ultralight aircrafts, similar to powered hang gliders, over the Wall back into East Berlin. While Holger circled overhead, distracting the guards, who were utterly confused when they saw a red star painted on the underbelly of the plane, Ingo landed and picked up Egbert and flew off, the three making it to safety in West Berlin.

Others tried to make their way through the waterways. In the Baltic Sea, one diving instructor managed to make it to Denmark using a brilliant homemade backpack-mounted propulsion device, turning himself into a mini human submarine, able to travel below the surface of the water without detection. Those trying to swim to freedom through Berlin's Spree River now found deadly metal spikes hidden below the surface.

By now prisons like the Hoheneck Castle were filled to the brim, packed with attempted escapees and nonconformists. In order to alleviate the overfill, in some cases prisoners were released, then kicked out of the country, forced to leave their children behind—the latter taken by the authorities and placed in state children's institutions or to be raised in a "proper communist household."

With a network now of nearly 200,000 civilian informants to spy on the East German population of 17 million, the Stasi perfected the sinister game of manipulation and control. While the Honecker administration worked to improve relations with the outside world, behind the Wall, Stasi psychological torture tactics increased. The secret police compiled dossiers on nearly everyone and used the information to hatch meticulously planned psychological attacks aimed at keeping people off balance: entrapping innocent people, then forcing them to turn on one another, creating deceptions to sow distrust.

Professional manipulators and master gaslighters, the secret police upped their use of ambient abuse to intimidate and terrorize, creating a murky cloud of mistrust that hung over the East. It's no wonder then that the goal for average citizens by this time was to try to live a life in peace, avoiding, if at all possible, coming up on the Stasi radar.

*I*n July 1975, Honecker surprised his countrymen again. In the presence of Soviet leader Leonid Brezhnev, U.S. president Gerald Ford, and West German chancellor Helmut Schmidt, Honecker signed the Helsinki Accords, providing East Germany with the international acceptance it craved. The agreement recognized East Germany's borders, but also called for respect for human rights and freedom of thought.

Shortly thereafter, in an astounding show of rapprochement with his people, Honecker announced "freedom of movement and travel" for the citizens of the East. This led immediately to hundreds of thousands of overwhelmingly excited East Germans applying for permission to emigrate. But the euphoria did not last when the great majority of applications were rejected. The exercise, however, was a valuable one for Honecker and the Stasi, for they could now finger those who had betrayed the country by expressing a desire to leave.

After years of conforming to the rules of a repressed society, people's lives had normalized; they didn't miss material things they didn't know existed or knew they simply couldn't have. Some, though, perhaps for the first time, felt a discontent rising among them and no longer believed what they were told. Some still wondered how much their circumstances really differed from their West German counterparts. Others secretly tuned in to the West for news, progressive fashion, and social trends at a time when it was in vogue in the

West to question authority. Some East Germans even began wearing their hair longer or sporting muttonchop sideburns and wearing bell-bottoms.

$\mathcal{U}$nder the constant gaze of the secret police, formation of an organized opposition was impossible. However, a few brave citizens risked everything to speak out against the regime in hopes of provoking political and social change. They fought alone for the rights of all citizens. The Stasi devoted an untold number of man-hours to keeping track of dissidents like Ulrike Poppe and Bärbel Bohley. Agents were posted around the clock to spy on them.

By car and on foot, the police followed dissidents relentlessly. Their telephones were tapped; video cameras and monitoring devices were hidden in homes, offices, anyplace they frequented, in order to catch them from an unimaginable array of angles. Even bedrooms were not off-limits, their private conversations and intimate moments recorded, then used to try to manipulate them and sabotage their activities. The Stasi investigated everyone dissidents came into contact with, including family members, classmates, colleagues, and friends, past and present. Anyone who associated with these so-called dangerous subversives ran the risk of a prison sentence, which essentially rendered dissidents completely marginalized from society.

With prisons packed to the breaking point, Honecker found a way to rid the country of "undesirables" and at the same time make money. He sold East Germans to the West.

In the West, human rights activists collected donations to try to free East German political prisoners; church groups pooled their money to liberate those oppressed for religious reasons; families wanting to get their relatives out amassed what they could. At $14,000 to $60,000 a head, hundreds of thousands were ransomed across for money or goods. Oil, copper, silver, fruit, corn, industrial diamonds, fertilizer, all were traded for people. The regime then sold the goods on the world market for hard currency. All told, earnings topped $1 billion, ill-gotten gains from a massive human trafficking scheme that ultimately helped East Germany avoid bankruptcy and kept the twisted regime afloat.

*I*n the United States, I was a fifteen-year-old high school sophomore. I longed for information about East Germany, but found little.

Time magazine ran an article about a defector who told of the complete degradation of East German society, of haunting human rights abuses. The most anyone seemed to really know was that East Germany was a harsh, bleak place, a strangely remote authoritarian communist state controlled by a ruthless secret police, and a staging ground for Soviet aggression into the West. What was clear was that, while some East Germans had managed to escape to the West and expose the hardships they had experienced, the full scale of oppression remained unknown as the great majority of stories of persecution and struggle in East Germany remained sealed inside, locked away and cut off from the rest of the world.

By the mid-1970s, my mother, Hanna, had been separated from her family for almost thirty years. Though she had struggled over the years to keep their memories alive, so much time had passed that, after a while, ache and longing gave way to a melancholy incompleteness, a hollow emptiness. She learned to live as best she could in the pain that she would never see them again.

Meanwhile, in the East, Roland had excelled as the director of his large school. Though he had done everything possible to earn promotion to school superintendent, he was informed that he had not been selected. The decision was due, his superiors said, to the fact that he came from a politically unreliable family. They specifically cited his connection to his sister who had defected to America, ironic since Roland had carefully avoided contact with my mother since her escape nearly thirty years earlier.

The superintendent job was given to a subordinate, a young hardline communist with an unblemished record. Roland had always understood that his family's checkered past in the eyes of the authorities could be used against him. Understanding that it was simply the way the regime operated, he had always half-expected it, and so he quickly adjusted, realizing there was nothing he could do.

But the matter greatly disappointed Opa and set him off on another tirade. The old bitterness, which seemed to have gone dormant, resurfaced when he saw that his oldest son, who had been a bright and shining star and a model example of the great teacher in the communist system since the day his career began, had been wronged. Opa felt a powerful new rage.

Around Klein Apenburg and in neighboring Apenburg, Opa started openly and loudly complaining about the regime and how it failed people. Clearly unconcerned whether the authorities were reading his outgoing mail, he sent Hanna a letter riddled with snide remarks and jabs at the regime, including this passage:

I am angered that the USSR and the USA can manage a joint space trip in the cosmos, but a visit of a U.S. citizen to Klein Apenburg

is not possible. Who can understand that? Maybe you should
make contact with the office of Erich Honecker to complain.

Once again, Oma implored him to keep his thoughts to himself.
She told Roland to speak to Opa about his resurging anger, which
he did.

"I'm seventy-seven years old," Opa replied. "What are they going
to do to an old man like me? They've already exiled me to Siberia."

The neighbors also warned Opa to pipe down. In Apenburg, store
clerks and shoppers kept their distance, not wanting to be associated
with his grumblings and antigovernment rants. Inevitably, he was reported.

The authorities came to Klein Apenburg. At his door, they
handed Opa a document. It was an official order committing him
to a "special program" at Uchtspringe, a hospital located fifty miles
away. Everyone knew of the place, recognized more commonly as the
insane asylum. The hospital in fact was a state psychiatric facility
and the special program was intensive reeducation training. Opa was
directed to go, as the order read, "to get your thoughts straight."

Not long after Opa was committed, on a cool September morning, Kai passed away.

16

A LIGHT SHINES
"OUR SOULS ARE FREE"
(1977)

The soul that sees beauty may sometimes walk alone.
—*Goethe*

*C*ordula was a happy, fresh-faced seven-year-old, the epitome of childhood innocence and purity. Fair-haired and fair-skinned like Reinhard, she was serene and levelheaded like him, too. With her blond hair cropped close and dimpled cheeks, she resembled a lovely little storybook pixie.

At her elementary school in Karl Marx City, along with reading, writing, mathematics, and learning about personal conduct in socialist society, Cordula and her classmates sang Russian jingles that made it easier to learn the language: "*Nina Nina tam kartina. Eto traktor i motor.*" (Nina, Nina, there is a picture. It's a tractor and an engine.) While she enjoyed school and Young Pioneer meetings, mostly she just loved to run, jump, and do anything physically challenging. It quickly became clear that Cordula had a gift for sports.

*W*hile most of the time East Germany was awash in a sterile gray, every October 7, like a bright red communist propaganda poster coming to life, the country ignited in a burst of color. On that day, whether they wanted to or not, East German citizens were called to celebrate "*der Tag der Republik,*" the Day of the Republic, the anniversary of the founding of their country.

In an over-the-top spectacle to showcase the might of the regime, every city and town throughout the country staged a huge celebration to mark the big day and call on citizens to take pride in their socialist communities. Modeled after similar celebrations in Moscow, spectacular pageantry hailed the regime, flaunted the power of the communist machine, and highlighted the progress of the nation. Workers, Party members, military and paramilitary units, and communist youth assembled to dazzle the crowds and galvanize the people to demonstrate their commitment to the state.

The bigger the town, the greater the show. In the capital city of Berlin, the fanfare was unmatched. Eastern European and communist world leaders were guests of honor as the regime showcased the latest military equipment supplied by Moscow.

Attendance was mandatory; every able-bodied citizen in every town and city was expected to show up and show his or her resolute support for the regime. Failure to attend or to demonstrate the required level of flag-waving enthusiasm was noted by Party leaders and secret police security interspersed throughout the crowd.

*I*n Karl Marx City, Heidi adjusted Cordula's red neckerchief, fastened a Young Pioneer lapel pin—a tiny metal badge in the shape of a flaming torch—and straightened the envelope hat on her head. Looking every bit the Young Pioneer poster child, seven-year-old Cordula patted her three-year-old sister Mari on the head and went off to take her place alongside her elementary school classmates who were already lining up for the parade.

In downtown Karl Marx City, crowds had already massed. Great banners lined the streets with reverberating messages: "Long Live Our Socialist Fatherland!" and "Workers of the World, Unite!" Red, black, and gold East German flags with the hammer-and-sickle coat of arms decorated the stands. Loudspeakers boomed defiant nationalistic music and up in the VIP box, local Party officials and honored guests looked down on the crowds, smiling, nodding, and waving. Nearby a group of effervescent youngsters held large, colorful bouquets of flowers, which they presented to benevolent-faced officials.

Heidi stood shoulder to shoulder with colleagues from her office. Someone handed out tiny flags, watching to see who did and who did not show an interest in taking one. Heidi thanked the man, flashing a convincing smile.

The parade began. Cordula marched past with the other fresh-faced Young Pioneers. Party cheerleaders, planted throughout the audience, made sure the crowds applauded at appropriate times and everyone knew by habit to follow their lead.

Genuine excitement surged when the athletes marched past. No one needed prompting to cheer them on. As in most countries, sports seemed to transcend politics. East Germans were particularly proud of their athletes and what they had been able to achieve to put East Germany on the map and bring honor to the country. As they marched past a roar went up, the crowds cheering as the athletes

waved back in return. And so it was in cities and towns throughout the East. Amid great displays of pomp and pageantry, parades
marked the country's progress and its great surge into the future.

*W*hile the rest of East Germany was in full-blown commemoration
mode, in Klein Apenburg all was deafeningly quiet. With two sons
deceased, a daughter three decades gone, a life of internal exile, and
a husband in an insane asylum, Oma's health began to take a turn
for the worse. Her children and grandchildren rallied around her in
Klein Apenburg, some even moving in for short periods just to be
with her and help with the chores. She put on a brave face every time
they came, but they saw that she had become fatigued. Seeing that
she had begun to neglect her garden, everyone who visited chipped
in to freshen the beds, pull weeds, and water the plants and flowers,
which had begun to wilt.

Later that autumn, Heidi and Reinhard, with Cordula and Mari in
tow, came to stay with Oma for a week. Still waiting after seven years
for the delivery of their new car, they came by train and then by bus.

Cordula could not wait to see what was growing in the garden,
including what was ripe and especially what could be eaten immediately. With Opa gone, she promptly took Oma by the hand to see the
vegetable beds, Cordula kneeling to pull up a gourd and a few green
beans. When the pain in Oma's legs became too much to bear, Cordula sat Oma in a padded chair that Reinhard brought outside and
placed on the edge of her garden.

Oma sat in that chair nearly the whole time that week, soaking up
the warmth of the sun, listening to the birds, and thinking back on
her life.

She reflected on the sustenance her gardens had always given her—lush greens that had seen countless rainfalls had at times given way to blanched and thirsty beds that had struggled through sporadic dry spells. Her greatest joy had been to toil until she saw the fruits of her labor, and especially when she saw everything blossoming in its fullest glory. It had been a labor of necessity to keep the family fed, but even more it had been a labor of love, and one that had come to nourish her and Opa in their isolated days in Klein Apenburg. Now, before the first frost arrived, it was once again time to prepare for the end of the season. Closing the chapter on any year had always made her melancholy as she canned and jarred and prepared for the winter. As Heidi picked the last of the season's vegetables, Oma sank into the sounds of Cordula's and Mari's joyous little-girl chirpings, which reinvigorated her spirit.

Over the next days, when the weather was mild, they relaxed outdoors, taking late-afternoon tea at a wooden table under an oak tree. Heidi baked fresh *pflaumkuchen* from the fruit she picked from Oma's plum trees while the girls frolicked, playing with the water pump or flitting around the yard like butterflies. While Heidi washed dishes and Reinhard cleaned garden tools or tinkered with Oma's bicycle, Oma sat in her chair, often with her eyes closed, breathing in the smell of the soil, and the aroma of late autumn in its fullest splendor.

In the evening they supped indoors, spreading a table with East German delicacies, including Eberswalder sausages with Bautzen mustard, and Spreewalder pickles that Heidi had found in the state shop. When night fell, they sat around the dinner table, playing cards and listening to Oma's favorite singer, Hans Albers, his silky voice crooning over the crackling radio. When Oma got tired, they helped her to bed, Cordula taking her time to tuck Oma in, patting her, and singing Sandman songs to soothe her to a peaceful slumber.

*T*hat week, Oma, reveling in a deep sense of peace and serenity, wrote a letter to Hanna that made it out to us in the States.

For the first time ever, she expressed herself openly, from the heart and without reservation, about her pride in her family, in Opa, in her children and grandchildren, and about the simple things that mattered to her. She wrote how she was "happy to have Heidi, Reinhard, and the girls visiting." She wrote of Manni and his family having visited the month before, and about the treats he had brought her: "glass cherries," plums, and apples. She wrote about Tiele's new teaching job, about Helga's vacation in Bulgaria, and about Tutti's darling blond-haired children. She wrote about Roland's achievements as a teacher, about his recent "retirement," and about her struggle with losing Kai and Klemens. She shared her grief over Opa's absence, even insinuating that the state had wronged him she took the risk to write, "it was something he did not deserve."

She ended the letter saying, "I find the greatest peace with my family and in my garden."

*O*n their last night in Klein Apenburg, Reinhard built a bonfire in the yard. Under a starry country sky, taking in the refulgence of the full moon on that clear autumn evening, they talked into the night, chattering animatedly about everything from the upcoming winter, predicted to be a bitterly cold one, to the amusing antics of the grandchildren. After a while, Oma became quieter. The silence lingered. When she spoke again, it was about a more serious matter: the future of the family.

"No one knows what the future holds," she said, "but I want the family to stay strong. Keep up the Family Wall. Lean on and protect one another no matter what may come." Though she had always avoided talk of repression in East Germany, Oma now had something to say about it. "No one can say what will happen or if things will change, but all I know is, justice will win. Truth will prevail and justice will win."

Silence lingered. After a time, she continued: "We have survived East Germany with our dignity intact. This life has not always been easy, but it has not made us bend. It has actually made us stronger. And we are strong because our souls are free."

No one spoke for a long time as they sat staring into the flames, watching them pop and crackle, little sparks snapping up and disappearing into the dark night sky. Heidi looked at her mother, her face illuminated by the light of the fire, her thoughts somewhere far away. Fine silver-white strands of hair, which were usually worked into a neat bun, now fell gently around her mellowed, age-lined face, deep creases framing her heavy-lidded eyes. Finally, she looked back at Heidi, the fire reflected in her eyes, a soft, knowing smile gracing her face.

All was quiet except for the crackling of the fire. After a long silence, Oma finally spoke again: "There will come a day when you will see her again. I may not live to see the day, but you will be reunited with Hanna."

On a cold, windless winter day, Opa was released from the insane asylum and returned home to Oma in Klein Apenburg. Over the next few months he said very little, and for as long as he lived, he would never talk about what had happened to him during those months at Uchtspringe. But his experience in the asylum marked the last time he would ever speak out against the East German regime.

Later that winter, Oma's health took a turn for the worse. By spring she had taken to her bed. With Oma having fewer good days than bad ones now, Opa tried to manage the household and tend to Oma as best he could.

Opa took to penning all their letters, writing what Oma dictated to him. At the end of one letter to Hanna he added a private note:

Mutti is doing so-so. She takes her medication regularly and goes to all her doctor's appointments. Is it possible for you to try to send some Sionon-diabetic sugar?

My mother did so, but she never got word that they received it.

By early spring, Oma could no longer ride her bicycle to the doctor in Apenburg. At the age of seventy-nine, Opa began making the runs for food and supplies. Since the doctor rarely made house calls to Klein Apenburg, eventually Oma was moved to a hospital in Apenburg.

All her children and grandchildren came to visit Oma. Cordula thought she seemed strong and full of hope, and was reassured when Oma promised, "I'll get well soon, and in the summertime you can come and play again in my garden." Then she turned to Heidi and said, "Keep the family together."

Not long thereafter, on the first of June 1978, on a warm spring day, two weeks before I graduated from high school, at the age of seventy-three Oma passed away with Opa by her side. The doctors said it was diabetes and high blood pressure. My mother got the news in the form of a Western Union telegram from Opa.

The family was devastated. The anchor of the family was gone.

She had been their center of gravity and the spirit that had sustained them. They were stunned sick, wondering how they could go on without her.

Years later, Hanna would write in her memoirs, "It was my father who opened my eyes to the wonders of the world and to the pursuit of knowledge, and it was my mother's spirit that has guided me throughout my life."

*O*pa was lost without Oma and took to spending his days on his "Opa's resting place" bench, staring the days away or sleeping on the couch instead of in their bed, or just lying awake. Roland came to take care of his father and they spent several weeks together grieving Oma's passing and trying to deal with their devastating loss. The families came to visit. Cordula sat with Opa, consoling him by reading and singing to him.

After Oma's death, Heidi occasionally looked over photographs of my mother and Eddie and the six of us children. By now, no one in the family really believed a reunion was possible, but everyone clung to Oma's hopeful words and her prophecy that they would one day see their sister Hanna again.

17

A SURPRISE FROM AMERICA
INNOCENCE
(1978–1980)

True innocence is ashamed of nothing.
—*Jean-Jacques Rousseau*

*O*nly one month after Oma died, one day in July, while everyone was still in mourning, the family was jolted by a surprise.

In the United States, I was due to start college in the fall. That summer, my older brother Albert, then an eighteen-year-old college sophomore, decided to take a summer backpacking trip through Europe with a high school friend. As he filled his pack in our kitchen at our home in suburban Washington, D.C., he casually asked our mother for Opa's phone number in East Germany. If he got a chance, he said, he would try to call Opa from West Germany.

For a couple of weeks the two teenagers explored Europe. In West Germany, Albert and his friend, with no real plan, boarded a train heading east and disembarked near an East German checkpoint. As expected, access into the country was tightly controlled, requiring foreigners to have submitted a detailed itinerary to the East German state tourist office up to nine weeks in advance. Western travelers were required to stay in designated hotels and register with the local police immediately on arrival. But at the security checkpoint, the two young, carefree, and adventurous college students, knowing nothing of the rules of entry into and travel through the country, submitted none of the required documents and simply dropped their passports into the slot.

The border guards behind the one-way mirrored glass, no doubt incredulous at the clearly naïve American teenagers wanting to travel alone in communist East Germany, did nothing in response, leaving the boys to wonder what could be taking so long. With armed guards standing nearby, the boys stood there waiting for twenty minutes until Albert's friend, whose father was an American diplomat, dropped a second, black diplomatic passport into the slot, to which the border guards immediately jumped into action, quickly processing the boys through.

The two boarded a train. Armed police with German shepherds moved through each car, inspecting it, then got off and the train moved eastward. The boys sat gazing out the window onto the drab gray countryside as the East German locomotive, loudly clanging and laboring, made its way at an excruciatingly slow speed through the countryside. An hour later, they disembarked at a station near Salzwedel. Instead of registering with the local police on arrival in the East as they were required to do, they simply walked into the desolate railway station and, in Albert's broken German, asked to use the telephone. The lone clerk, startled by the sudden appearance of a couple of foreigners, obliged and Albert called Opa's telephone number.

Roland answered the phone. Stupefied, he only managed to say,

"Ich komme so fort!" (I am coming to you now), before quickly hanging up. Then he dialed Manni, telling him to "gather the family immediately," with no further explanation given over the likely Stasi-monitored phone line, leaving Manni to believe that their seventy-nine-year-old father was ill. Then Roland grabbed his car keys, ran out the door, and jumped into his car, his heartbeat racing. Not wanting to draw attention from the neighbors, he took a deep breath, tried to calm himself, and slowly backed out of the driveway. Once on the open road, he accelerated as fast as his tiny Trabant could go, smoky exhaust trailing in a billow behind him. Nearly an hour later he reached the train station. Upon seeing the two boys, Roland stopped and emerged from the car. He and Albert embraced, Roland looked around, and they all quickly got into the car and sped off.

By then, as many of the family members as could make it on such short notice had assembled at Opa's place, and more would stream in throughout the day. Roland parked behind the house, and he and the boys entered through the side door so they could not be seen by nosy neighbors who would surely be suspicious of visitors in University of California T-shirts and American blue jeans.

Once inside the house, Roland closed the door, fell back against it, and breathed a sigh of relief. The family stood up slowly, gazing at the two teenagers.

"This is Albert," Roland said. "Hanna's son from America."

They were in utter shock. They stood rooted in place, staring at Albert as if he were an alien from outer space. How was it possible that their long-gone sister's son had managed to slip into the country, avoid the attention of the authorities, and was now standing among

them in Opa's living room in Klein Apenburg? Tutti let out an ecstatic wail and launched forward to grab Albert by the face, then
pulled him to her, hugging him tightly.

Suddenly everyone was swarming Albert, overwhelmed with
emotion, tears streaming down their faces as they reached to embrace him: Tutti's daughters and her husband, Manni, his wife and
children, Helga and her son. It did not matter that Albert spoke only
kitchen-table German and the family spoke virtually no English;
tears and laughter, animated, heartfelt gestures, and expressions of
great elation and deep emotion carried the conversation. Albert's
open personality and easy smile seamlessly connected with Manni's
lighthearted way, Helga's doting motherly pats, Tutti's rugged openness. The fact that they had never met seemed irrelevant.

After a few minutes, Roland quipped about Albert being smothered and suggested they back off and give him some breathing room.
Still clinging to him, they retreated slightly, and the huddle parted.

Then Albert saw him standing a few feet away. He broke from the
group and went to our grandfather. Opa, his nerves bleached by years of
punishment, a man not prone to physical contact, embraced his grandson.

The next day, they took Albert and his friend out the back way, for a
walk in the countryside, using a circuitous route so they could avoid
being seen by the prying eyes of neighbors, which befuddled the boys,
who had no understanding of why they wound so strangely around instead of walking directly to the nearby field. After a pleasant walk, they
returned to the house along a different but similarly roundabout route.

Later that day, they whisked Albert to the cemetery to visit Oma's
grave, the soil freshly turned from her recent burial. The boys stayed

for nearly two days. On their last day, with long embraces and plenty of tears, the family said good-bye.

Roland drove the boys back to the train station, along the way wiping his eyes and imparting words that Albert barely understood but could tell included heartfelt greetings and loving sentiments for Hanna. A short distance from the station, Roland let the boys out and quickly drove off. The two promptly bought a ticket and boarded the train. Close to the border, the train stopped and armed East German police with dogs inspected the cars and checked everyone's papers. The boys rode to the border and were quietly processed out of East Germany and back into the West.

My mother knew nothing of her son's travel to the family in the East until Albert returned to the United States two weeks later and told her about it. Shocked to learn what he had done, she paused to let it sink in. Then she praised him for his daring, and asked him to tell her all about her father, her brothers and sisters, nieces and nephews, how each of them looked, what they did, what they said. She reveled in every detail and, some days later, lost herself in the photographs he developed and brought to her. (See page 8 of photo insert.)

By the late 1970s, the East German sports program was the pride of the nation. To the great surprise of the international sports community, East German athletes began racking up remarkable results in world competitions, including at the Olympics. In the late 1970s and on into the early 1980s, the tiny country of East Germany came in second in overall medals count in three consecutive Summer Olympics, just behind the far larger Soviet Union, and well ahead of a stunned West Germany. The world looked on in awe as East Germany churned out an unbelievable cast of champions, dominating a number of sports, including swimming, gymnastics, and cycling.

Honecker was pleased. His calculated quest to prove to the world that East Germany deserved international prestige, recognition, and respect was really starting to pay off.

Fueled by their successes, sports authorities continued to search out the most physically gifted young people with the potential to be molded into the best athletes in the world. Thousands of talent scouts and physical fitness instructors, coaches, and trainers were brought on board to scour the country to find those they believed would be most profitable to the program and set them on the path to becoming world-class athletes.

Determined to discover new ways to push the human body to new heights, the sports program created and developed cutting-edge, science-based methods they believed would give East German athletes a competitive advantage. Trainers were required to have a keen understanding of anatomy, psychology, physiology, and biomechanics, and to be master tacticians in recognizing optimal strain loads, channeling each athlete's mental focus, and maximizing their performance potential. Top sports medicine doctors and research scientists created pioneering theories and techniques at the Institute for Applied Training Science, the sports research headquarters, and amassed a 75,000-volume sports library dedicated to the minutiae of sports mechanics. Advanced for their day, East German researchers exhaustively studied their competition from every angle, making video recordings of all the details of each major competition so they could analyze every potential opponent's strategy and create new tactics to counter them.

In the coming years the state would invest hundreds of millions of dollars to pursue its goal of dominating the world of sports. Besides state security and funding to maintain the Wall, East Germany's sports program held the biggest slice of the national budget.

*I*n Karl Marx City, Cordula was already proving to be a dynamic addition to her school and Young Pioneer group, especially in her athletic ability. At youth events and sports activities, heads turned when she ran, climbed, or swam. She loved almost every sport she tried and was good at all of them. She had remarkable speed and agility and easily outdid most of the boys. Awed by her natural physical ability, her fitness teacher promptly informed national talent scouts.

They came to observe her in the pool. She was strong, they noticed, and had focus, intensity, and remarkable natural talent. She glided through the water like a fish. The scouts invited her to compete in swim trials. At tryouts she impressed them, showing remarkable promise for an eight-year-old. It was clear to the trainers that she had incredible potential.

Cordula easily passed the physical test, but she would also have to pass an assessment to gauge, among other things, whether or not she might have the potential to be a flight risk should she rise through the ranks and someday be chosen to travel outside East Germany. The sports program simply could not afford to recruit potential defectors and they would carefully monitor and vet each and every athlete along the way. A defection would come back to haunt the recruiter who had promoted the athlete who escaped; the incident would be a huge embarrassment to a regime that was trying so hard to impress the world.

Heidi assumed Cordula would be disqualified during the vetting process once her interviewers learned about Cordula's family black marks. With Heidi and Reinhard sidestepping membership in the Communist Party, Opa's numerous run-ins with the regime, and Hanna's flight from the republic and marriage to a U.S. Army officer, it hardly seemed that Cordula would have a chance.

*I*n the administrative offices of the Karl Marx City sports complex, with her mother waiting in the hallway, Cordula knocked and entered the room alone. She was directed to take a seat. The officials looked her over. Then one of them gently asked, "Would you like to travel to America to see your aunt?"

She lifted her head and looked at him. "No."

"Do you love your parents?"

"Yes."

Satisfied, he smiled. They congratulated Cordula and instructed her mother to withdraw her daughter immediately from her regular school in order to enter a specialized sports school to begin intensive training.

At her new school, training and competition took priority over academics. Runners ran, gymnasts tumbled, swimmers swam. Within a year, Cordula was already winning at local and regional junior competitions. At nine years old, she was selected for the East German national junior swim team, joining the ranks of the country's finest young female swimmers.

*A*s Cordula advanced into the junior swim program, I was a freshman at James Madison University, in Harrisonburg, Virginia. I studied Russian and in history class I learned more about the Cold War, about escalating U.S. and Soviet tensions, battles between communism and democracy around the world, and the dangers of

Soviet communist influence. By the late 1970s, détente had fizzled out and superpower tensions had returned. Nuclear confrontation was once again a real threat.

*O*nce again, letters from the East to the United States slowed to only a few a year. Few made it out, and few made it in, and my mother, Hanna, had to face the reality that the family had once again disappeared from her.

Because there was still so little firsthand information about what was going on behind the Iron Curtain, stereotypes abounded about godless, brainwashed, automaton-like trapped souls with a near-alien ideology. The closest we got to any definitive information about the people in the East was through the eyes and words of Boris Pasternak in *Doctor Zhivago* and Aleksandr Solzhenitsyn's *The Gulag Archipelago*, books smuggled out of the Soviet Union.

The world's most reclusive country, at best, we knew it was a repressive police state with an appalling human rights record, and that it had extraordinary athletes.

*A*s Cordula trained among the best young athletes in East Germany, she was getting her own education about a much-revered and powerful Soviet Union and, in complete contrast, a hostile and corrupt United States. Propaganda films drilled into people's heads the notion that America was a diseased country, rampant with crazies, criminals, and unemployed, destitute citizens. It was, she was told, an economic disaster zone and NATO a raging force of great destruction. East Germany, on the other hand, was a loving, peaceful nation, with a government that truly cared about its people and provided them everything they needed, and had built a safe and secure environment far removed from all the dangers of delinquency and chaos.

*T*op East German Olympic athletes trained on state-of-the-art equipment imported from the West, had the best in sports attire, and enjoyed many perks and privileges unavailable to the rest of the population. Those just entering the system at the junior level, however, had fewer resources and had to earn their way up. They trained in regular gyms, community sports centers, and in local *Schwimmbads*, community swimming pools, using whatever was available.

Every day at swim training, Cordula donned her standard-issue navy blue swimsuit. After months of use, the poor-quality material became thin and transparent; when they were wet they stretched out and sagged awkwardly on the little athletes' bodies. Eventually the girls' chests and bottoms were exposed. When the swimsuits reached their expiration for wearability, buttons were simply sewn onto the shoulders to keep them from slipping off completely.

One day, my mother received a letter from Heidi saying that the family was fine. Mari had entered school; not wanting to jeopardize Cordula's standing in the East German sports program, Heidi wrote simply that Cordula liked to swim. Several months later a package arrived from America, postmarked from West Germany: a new dress for Mari, and for Cordula a fancy new swimsuit that Hanna had bought at the Fort Myer Army Post Exchange, just outside of Washington, D.C.

To Cordula everything about the American swimsuit was exotic and beautifully different from anything available in East Germany. With a mod, flower-power design, it was colorful: pink, orange, and yellow on a striped background and, best of all, it was made of high-quality fabric that held its shape. Cordula was overjoyed.

Cordula continued to wear her saggy blue suit to practice but when she came home, she put on her American swimsuit and privately admired herself in the mirror, twisting to see from behind, from the front, how it looked when she dove from the side. At practice she continued to excel and before long had earned a bronze medal in her first national junior competition. Emboldened by her win, she toyed with the idea of wearing her American swimsuit to practice. One day, desire won out. In the locker room, she changed into her swimsuit and proudly walked out to the pool and joined the other girls.

Her trainers were distracted, too busy hopping around trying to accommodate a visiting East German media team, so, to her delight, no one seemed to focus any attention on her. But suddenly a trainer called out to Cordula. She swallowed hard and ran over, landing in a gymnast's stick in front of him. Lowering her gaze, she prepared to be admonished for wearing such unorthodox attire that clearly did not come from East Germany instead of wearing the standard team suit. But he was in a hurry and did not even look up at her. Several other girls were called out and all were told to hastily retrieve their medals and immediately report to the camera crew outside. An official medal-winner team photograph was taken with Cordula standing alongside her teammates while wearing her American swimsuit, along with an innocent, slightly impish expression, feeling pretty and proud. The photo ran in the newspaper the next day alongside a feature story about up-and-coming superstar East German athletes. (See page 9 of photo insert.)

In the late 1970s, the two superpowers resumed feuding on a major scale. In 1979, the Soviets invaded Afghanistan, and in protest, the next summer, America boycotted the Summer Olympics in Moscow.

*B*y now, having racked up $10 billion in foreign debt to Western creditors, East Germany's economy was in shambles and slipping into bankruptcy. As suddenly as consumer supplies had appeared in the 1970s, by the end of the decade much had disappeared from store shelves and the quality of products seemed to be going from bad to worse. What was available became rationed or marked up, often becoming prohibitively expensive, the meager wages of the average worker preventing them from being able to purchase even these inferior products.

East Germany had peaked and the system was no longer improving. In fact, things seemed to be going in the opposite direction. Nevertheless, as usual, Honecker reassured his people that the country was well respected and seen as a leader not only in the East Bloc but also in the West. At home and abroad, Honecker pointed to the country's sports achievements as proof of progress. East German athletes were setting new world records and catapulting the reputation of the country to new heights, enabling East Germany to finally gain legitimacy from the rest of the world, which, he said, it so richly deserved.

To those who for years had held on to the belief that the country was making substantial progress, it had become evident that it actually wasn't. Though many had suspected or known this all along, others were beginning to awaken from a thirty-year slumber to see the truth: a decades-old promise by Germany's now-aging dictatorship would never be fulfilled.

*B*y now 120 people had died, either shot or drowned, trying to get out of East Germany since the building of the Berlin Wall.

Then in 1979, a spectacular incident occurred when two families took their escape to the skies. Having fashioned a hot-air balloon from canvas, bedsheets, old scraps of fabric, and a homemade gas burner, Günter Wetzel, a mason, and Peter Strelzyk, a mechanic, and their families ascended into the dark night sky and sailed quietly over the Wall to safety in the West.

The escape made headlines around the world, with Strelzyk saying, "Freedom is the most valuable thing a human being can possess. The only people who know that are people who have had to live without it. If you've grown up free, you don't know what it means." After that escape, the sale of fabric and cloth was closely controlled in East Germany.

But far more deaths occurred at the border than escapes. Not long after the successful balloon escape, in two separate incidents, eighteen-year-old Marienetta Jirkowsky and Dr. Johannes Muschol were shot dead at the Wall; the lifeless body of thirty-two-year-old Peter Grohganz was pulled from the Spree River; and twenty-six-year-old Thomas Taubmann was killed leaping from a bridge while trying to make a break for the West.

*E*ast Germans started craving increased information and contact with the outside world and it became an open secret that many of them were tuning in to the West. While some chose not to listen to what the regime said were lies, and some were still afraid to tune in for fear of being reported by patriotic neighbors, many others started to tap into the airwaves.

With renewed tensions, East Germany once again tried to jam BBC and U.S.-sponsored broadcasts like the Voice of America, which brought pro-Western messages about freedom and democracy to the East.

Border areas logically offered the best reception while some regions had little or no access to the airwaves due to distance or geography. In Karl Marx City, Heidi and Reinhard were able to tune in and did so. The reception in their flat was weak but they had relatively good access to West Germany's ARD and ZDF TV stations as well as to Radio Luxembourg and Deutschlandfunk, which became their favorite source of news and music. Reinhard's parents in Stollberg tuned in to Tagesschau, a West German news service, but relatives living in nearby Dresden were unable to receive the signal. On occasion, Heidi was able to pick up grainy reception of the American soap opera *Dallas*. She marveled at the lavish American lifestyle of champagne, jewels, and Cadillacs; did most Americans really live like that? The regime's worst nightmare, this mass movement of behind-closed-doors East German civil disobedience would grow and eventually become known as "defection by television."

*O*ne day Reinhard was dispatched to his manager's apartment to fix an electrical problem and found the boss's daughter glued to the TV set, happily watching a West German children's show despite the boss's own earlier strict order to employees not to tune in. What was clear was that, by now, the regime knew it could not possibly track every instance of eyes and ears tuning in to the West.

To counter the television and radio threat from the West, the East German authorities aired *Der Schwarze Kanal*, the Black Channel, a play on words meaning sewer flow, a weekly propaganda show designed to discredit West German news stories. On Monday evenings at nine, East Germans could watch heavily edited versions of West German newscasts twisted with East German overdubs and political commentary meant to confuse and incite people to question the West's version of the truth. However, *Der Schwarze Kanal* gained a reputation as the least popular show on East German television.

As if Western television and radio broadcasts weren't enough to cause concern in the regime, there was another reason for alarm: music.

In the United States, I was a typical American teenager of the '70s, with long, carefree hair and dreams of world peace. I wore bell-bottom jeans and platform shoes and listened to Cat Stevens and James Taylor. They sang of love, peace, and freedom—things important to young people all over the world in that era.

I loved the Beatles, as did millions in both the West and behind the Iron Curtain, where they had taken on near prophet-like status, even years after they had disbanded. Communist authorities perceived the Beatles' music to be a grave threat—a countercultural phenomenon that spread antigovernment messages of revolution and questioning authority; just as dangerous, the Beatles' lyrics of love and peace threatened to humanize the West. The East German authorities awkwardly tried to neutralize the Beatles' impact by publicly branding them as deviants and referring to them as "a horde of wild mushroom heads," but any attempt to lessen their influence only encouraged young East Germans to try to listen to them even more.

In an attempt to counter the Western music threat, the leadership tried to appeal to East German youth by offering its own brand of regime-friendly pop music. *Ostrock,* or East Rock, bands like the Puhdys and Silly were strictly monitored to make sure they didn't step out of line or weave double meanings into their songs, and when they played too close to the edge of what was allowed, the authorities simply put an end to their music.

One of East Germany's music superstars was my distant cousin Erich Klaus. Because of his Communist Party loyalty and willingness to sing songs sanctioned by the regime, he enjoyed spectacular celebrity, which came with plenty of perks, including access to Western goods, foreign clothing and shoes, a luxury apartment, and a lot of travel abroad. Because of his special status, he distanced himself from common citizens, including his own relatives, especially those with black marks like my mother's family, believing, probably rightly so, that association with such a family would only jeopardize his privileged position.

In one letter to my mother Opa wrote, "Erich Klaus is now world famous. We hear him all the time on the radio. He gets to travel a lot, was just in Mexico and will soon go to Austria. He does not keep in touch at all."

18

PARADISE BUNGALOW
REFUGE AND SOLACE
(1980–1982)

The only way to deal with an unfree world is to
become so absolutely free that your very existence
is the act of rebellion.
—*Albert Camus*

*B*y the early 1980s, America had a new president. Ronald
Reagan vehemently denounced communism and warned against ig-
noring Moscow's dangerous ambitions. Americans steadfastly supported their new president's impassioned anticommunist stance and
continued to be hyperalert about Soviet spies and communists in
their midst.

In Washington, D.C., my parents attended a gala of the Association of Former Intelligence Officers. Beneath a sparkling chandelier, amid tables covered in white linen, flowered centerpieces, and
stemmed glassware, a local politician's wife, looking to make polite
dinner conversation, pleasantly asked my mother where she was from.

"East Germany," my mother gracefully replied, to which the alarmed woman gasped, "You're a communist?!" That outburst brought the table to a sudden silence. After that, whenever my mother was asked where she was from, she lied and said she was from Hannover, West Germany.

By now East German authorities realized they simply could not adequately feed their people, so the regime gave out 850,000 plots of land to citizens to use to grow their own food. Allotment gardens would allow weekend farmers to grow their own vegetables and fruit in order to feed their families and help stock state stores in order to sustain others. The move was a stinging defeat for communism, which was supposed to raise enough food on the collectives for the masses. But in East Germany that concept had failed, so the regime was forced to come up with a new plan. And so they reinstituted individual ownership of land. This was semiprivatization in a country that had outlawed private enterprise.

At work, Heidi's boss, Meier, assembled his workers and read from an order that the state was rewarding them with a small plot of land they could use to harvest their own food. The workers were thrilled; there were smiles all around. Naturally, Communist Party members were to receive priority as there were a limited number of plots, and not everyone would be selected. Heidi stood in the back. She knew she would be at the bottom of the list and did not hold out much hope of receiving such a highly prized gift from the state.

A similar proclamation was made at Reinhard's workshop, where he too looked on as Party members were called forward. For the naturally industrious Reinhard, it was a fantasy to believe that he could essentially have his own mini-farm.

Several months went by when Heidi was called to Meier's office, where, out of earshot of others, he recognized her for her outstanding performance. With a reserved smile, he handed her an envelope containing an announcement that she had been selected to receive a plot.

She raced home that evening to report the news to Reinhard. Incredulous at their good fortune, the two embraced, shedding tears. Reinhard read the notice over and over, studying every word. Nothing could be better, he thought, than owning a piece of land in the countryside where they could not only grow their own food, but even more important, have a sense of independence, ownership, and freedom. This was a windfall, he thought, and he fully intended to run with it.

That night, Reinhard, who by not joining the Communist Party had hampered his opportunity for education, went to bed with his head spinning with ideas about how to best design and build not only a bountiful garden, but a dwelling of some kind, a little country home, where they could even stay overnight on the weekends.

That first weekend after receiving the envelope, Reinhard, Heidi, Cordula, and Mari took a bus, then walked the nearly two miles to claim their little patch of land. On the outskirts of a forest they came upon a large clearing about the size of four football fields, with fertile, unturned soil. Reinhard canvassed the grounds, then carefully studied the map provided with the paperwork, finding their allocated lot on the south side, near the middle of the field. There, one foot in front of the other, he paced out the perimeter of their land, a 24x18-yard rectangle, mindfully measuring and marking every inch of the circumference of their plot, one of some forty that would become gardens on that undeveloped stretch of land.

Over the next few weeks, other citizen farmers started arriving to claim their new land as well. Before long there was a frenzy of activity, happy fellow landowners digging and raking to loosen the soil, as the earth was prepared for planting, everyone sharing tools, wheelbarrows, helping one another to get their gardens up and running. Soon people were passing along gardening techniques, exchanging tips about where to find the best materials, shovels, rakes, and seeds.

This was the kind of zeal and passion the leaders of East Germany had envisioned would propel the socialist country to new heights, but which had never materialized. Yet when provided the means to do so, these hundreds of private small-plot farmers, who controlled only a tiny fraction of the county's arable land, would in the coming years produce almost half of East Germany's vegetables. It was a dilemma for the authorities, who worried about the long-term impact individual enterprise could have in undermining their command-and-control economy.

It also greatly concerned the regime that the garden plots could be seen as a refuge from the constraints of communist society.

Over the next weeks and months, Heidi and Reinhard spent every spare moment working on their land, and after a while their crops began to bud. Then Reinhard acquired some concrete, which he mixed and poured into a 12x25-foot foundation. For the equivalent of about sixty dollars, he bought a prefabricated shed kit, a box of crude parts that included little more than four thin, poorly constructed wooden plank walls, a cheaply made door, and a quarter-inch plywood roof. Taking the challenge to a new level, with paper and pencil he reconfigured the layout, switched out and recast parts, improvising at every turn to improve the shack's design. Along with his architectural drawings, they hauled the parts to the plot.

For Reinhard, the search for materials and building supplies for the cabin turned into a near obsession, his every waking moment preoccupied with plans about how he could best improve the family's new country retreat, how he could use wooden planks he had scrounged to build braces to strengthen the cabin's frame, or repurpose a corrugated metal fence into the ideal weatherproof roof.

Every Saturday morning, except for weekends when Cordula had swim competitions, the young family quietly slipped out of town on a bus, loaded down with tools, new seeds and wood planks, glass panes, salvaged bricks, all sorts of materials they would recycle into some useful part, odds and ends they were able to scrape together, had scavenged, traded for, or bought. While their parents turned the soil, creating one entire section just for flowers and others for vegetables and berries, the girls played nearby in the dirt, watered the plants, or dropped seeds into holes. At the site, Heidi and Reinhard worked indefatigably from sunup to sundown improving their cottage and garden, often finding themselves still working long after other gardeners had gone home. Exhausted at the end of the day, they often slept under the stars in a makeshift tent, even when it rained.

Bit by bit, the bungalow came together. These were immensely happy days for the family and they reveled in their work and in every little achievement, in every new construction and every tiny bud or shoot that sprouted. In just a few seasons, their square of land had been completely transformed.

In the summer of 1982, two years after they had first received their dirt plot, they were tremendously proud of what they had achieved. Together the four stood standing, taking it all in.

Surrounding their plot was a tidy chain-link fence with a wooden picket gate. Inside the gate was a lush wonderland carpeted in a blanket of green grass. In the garden beds, yellow and green pepper plants stood tall, and leafy growth from celeriac, kohlrabi, and beets shot up through the earth. Tomatoes climbed upward along wooden dowels and on the garden floor, parsley, chives, beans, and zucchini grew and ripened. Farther in, tiny pink buds dotted dark green vines that curled around a homemade wooden arbor, and red flowers peeked out from an old, broken-down wheelbarrow-turned-planter.

In the middle of it all, standing resolute against the blue sky, was the cottage. A small miracle in the form of a handsomely built, modest country cabin, though hardly luxurious, it had an enchanting, dreamlike quality and bore all the excitement, magic, and innocence of a child's secret fort tucked away in the forest.

They took in a breath. Reinhard took Heidi's hand and together they walked up the stone-brick path and block concrete steps to the wood-planked porch and into the cabin. Inside to the left stood a tiny makeshift kitchen complete with running water sourced by a large barrel of rainwater.

While their flat in the city remained sparsely decorated, they had adorned the bungalow with expensive Plauen lace curtains that now graced the windows of the main living space, a six-by-eight-foot sitting room with wooden cabinets that Reinhard had fashioned from wood scraps and fastened onto the wall.

Below the cabinets stood a small couch made up of patched pillows and sofa parts propped up on a wooden frame. Three chairs they had hauled from the apartment stood on either side of a card table. In the center of the table, a drinking glass with fresh flowers topped a white, handmade lace doily and a small battery-powered radio.

Around the corner from the living room was a small alcove with a fully functioning dry-composting toilet. Across from it there was just enough room to squeeze in a narrow double-decked sleeping loft. Thick, fluffy comforters topped two wooden beds. They went back outside and turned to face the bungalow.

It was, to put it simply, a triumph—a testament to ingenuity and perseverance in a society that forbade individuality and independence and shunned innovation and creativity on one's own terms. The natural-born engineer who had not been given the same chance as others had found a way to achieve something out of nothing.

A victory not just in the physical sense, their little paradise stood as a symbol against the limitations imposed by communism, against all that their government had denied its people, and in fact against everything the regime stood for. They had created a place where they could retreat to not only their own private physical space but, along with it, their own private thoughts. Despite being born into and living their entire lives under the haze of authoritarianism, at the bungalow they felt free.

As night fell, they lit a candle and put the girls to bed, then Heidi and Reinhard returned outside to the darkness, where they sat back and uncorked a bottle of Rotkäppchen sparkling champagne they had bought and saved just for this one glorious moment.

They toasted to each other and to their life together. Then Reinhard turned on the radio. After searching for a while, he picked up a scratchy signal from the West, tickled when he happened upon a singer he had recently discovered: a man by the name of Elvis.

Smiling, they listened to the radio eking out fuzzy electronic squelches with a barely audible but definite "Love Me Tender."

With no one around within earshot, they sipped champagne into the night, talking and panning the cloudless sky, which seemed to twinkle with a thousand silver stars as they reveled in their tiny slice of paradise and freedom in the middle of a country that was not free.

It would be the first of many overnights in the bungalow, nights in which Heidi would always sleep deeply, her dreams often filled with themes of flying, of soaring like a bird in the vast open sky, looking down over the land and then flying over the border and onward into the horizon, free to soar wherever she wanted, to places unknown and undiscovered.

Occasionally Oma came to Heidi in her dreams, her paeans of wisdom resonating: "Our souls are free. Keep up the Family Wall. Take care of each other. Justice will win. Someday you will see Hanna again. . . ."

Throughout East Germany, more and more citizens were tuning in to the West. Sitting up close to radios and televisions turned down low, East Germans began to learn the truth about the growing Solidarity movement in Poland, about life in West Germany, as well as listening to music from the West. They marveled at Western innovations like VCRs and Walkmans, becoming aware that West German products, electronics, appliances, cars, and clothing were of far better quality; East German versions were not only inferior but also prohibitively expensive as well. Even though the regime had told them otherwise, that Western products were expensive, poor in quality, simply unnecessary, and contributed to the delinquency of people, they could see for themselves that it was not true.

Little by little, though the regime continued to insist its citizens had everything they needed, people began to awaken—to piece together the truth about the world outside and about life in the West,

Reinhard breaks ground where he will build the bungalow.

which, they came to realize, was far less menacing than they had been led to believe. But for many, there was still a lingering fear and almost an innocence of the unknown that lay on the other side of the Wall.

But the truth was, beyond the empty shelves and poor selection, Heidi and Reinhard, like millions of other East Germans, did not really care about material things. It was really only freedom that they longed for.

PART FOUR

19

ASSIGNMENT: BERLIN
INTELLIGENCE OPERATIONS
(1982–1984)

The West won't contain communism. It will
transcend communism. We will dismiss it as
some bizarre chapter in human history whose
last pages are even now being written.
—*President Ronald Reagan*

I graduated from college in 1982 and was commissioned a second lieutenant in the U.S. Army as an intelligence officer. I chose the army in part because I came from a family that wanted to pay back, who loved what America stood for and believed in service to a country that was built on the principles of freedom and opportunity. I also wanted to do something different with my life, and saw a career in intelligence as an interesting way to try to make a difference.

At around the same time I was raising my right hand to take the oath to "support and defend the Constitution of the United

States against all enemies, foreign and domestic," in Karl Marx City, Cordula was taking an entirely different kind of oath. A teenager entering her adult years, she was raising her right hand to take the Jugendweihe promise to dedicate her life to serving the East German regime. Since her mother, Heidi, had taken the oath back in 1964, the words to the pledge had been changed and now had a much more militant tone, with the words "embracing peace-loving people" replaced by "to defend socialism against every imperialist attack."

Cordula continued to train alongside other new recruits in the East German junior swimming program. Heidi and Reinhard worked at their jobs during the week and relished their weekends at the bungalow, constantly improving and beautifying their private little world, which grew more lush as the years went on.

Roland, Tiele, Manni, Helga, and Tutti's families went about their lives, working, raising their children as best they could, taking vacations inside East Germany when they could, to the Arendsee lake, up to the Baltic coast or down to the Thuringian Forest.

Opa spent his days quietly in Klein Apenburg, much of the time on his "Opa's resting place" bench. At the age of eighty-three, when he could no longer care for himself, he was placed in a retirement home in Poppau. Once there, he asked Roland to get him a tape recorder so that he could "chronicle the family history."

He filled these tapes with details about his ancestors, talked about German history and philosophy and recited poetry. He mentioned nothing about his struggles under the regime, choosing instead to speak about things that had made him happy during his lifetime, especially his wife, his children, and his grandchildren. Perhaps most remarkably, he remembered with the greatest fondness the students he had taught over the years, recalling with delight and tenderness their amusing antics, such as when "six-year-old Karl Schinkel had found a baby sparrow and wanted to put him in a box and teach it how to sing and lay eggs."

After college graduation, I reported for intelligence training.

It was the first time I had ever seen the desert. I was mesmerized by the open sky, its vast, wide-angle beauty. On a hot day on a remote Texas highway, the road ahead seemed endless.

Truckers honked and waved as I passed by, a carefree young girl, with the windows rolled down, singing along to the radio in a ten-year-old blue Ford Granada with Virginia license plates, the old family car my parents had given me for my trip west. It was the first step of my journey into adulthood. I was now some sixteen hundred miles from home and on my third day of traveling cross-country alone. Somewhere on the long stretch of mostly uninhabited road between Amarillo, Texas, and Albuquerque, New Mexico, my long hair blowing wildly about in the breeze, I sang along, aimlessly and with abandon, to country songs I barely knew as I barreled westward down Interstate 40.

Carefree and independent on the wide-open rural Texas highway was a liberating place for a twenty-one-year-old to be. With an abundance of youthful optimism, I was excited to be heading off into the world to face whatever adventures came my way. The endless sky, the never-ending horizon, and the mystery and allure of the desert seemed a perfect welcome to the great unknown that lay ahead.

Times were changing for women in the military. Though many men were still uneasy about it, some doors were opening and women were starting to play more of a role in the army. In 1973, the first women graduated from U.S. Army Airborne School and in the early 1980s, the first female cadets graduated from the U.S. Military Academy at West Point. The Women's Army Corps, the WACs, was disbanded and female soldiers and officers were allowed to join the ranks of the regular army, serving in a limited number of career fields. Luckily for me, intelligence was one of them. Women were still restricted, however, from serving in combat or on enemy soil.

After nine months of intelligence training at Fort Huachuca, Arizona, I was sent to Airborne School at Fort Benning, Georgia, where I was one of three women in my Airborne Training class of more than 250 aspiring paratroopers.

Training was led by tough, seasoned instructors called Black Hats, whose job it was to set a high standard and weed out trainees who weren't mentally or physically up to the task. As the class's female officer, I was constantly in the Black Hats' crosshairs. They put me in the spotlight, seemed to enjoy testing me relentlessly, making me do extra push-ups when any of my fellow trainees were found guilty of any tiny infraction. They put me in a leadership position as first squad leader, first planeload commander, and gave me the "privilege" of being the first to jump from the airplane every time. After three weeks of training, I graduated. My brother Albert, by then an army captain, and my father, who wore his uniform, came to "pin on my wings" at the graduation ceremony.

After Jump School, I was given my first assignment: Berlin, deep inside East Germany.

My family, especially my mother, was shocked by my assignment, given that I had reported my East German relatives on my security forms. Eddie, my father, advised that if anyone from the East should ever try to make contact with me, I was to report it to my superiors immediately. If foreign intelligence, including the Stasi, approached me, threatening to harm the family in the East if I did not comply with their demands, in an effort to shut down any threat to the family I was to convincingly respond, "I don't know these people. They mean nothing to me."

In the fall of 1983, I flew to Berlin on a Lufthansa commercial flight. After passing over West Germany, our aircraft entered East German airspace, flying through one of the three prescribed twenty-mile-wide, 10,000-feet-altitude flight corridors required for civil and Allied military air traffic to and from Berlin.

As we flew over the Inner German border, down below lay the heavily fortified 860-mile divide that separated the two halves of Germany, its length running from the Baltic Sea in the north to Czechoslovakia in the south. On the west side of the border, vigilant German and Allied forces patrolled, guarding against a Soviet invasion.

During the flight, many thoughts raced through my mind as I wondered what awaited me when I landed. I tried to focus on the West Berlin guidebook on my lap, but found it hard to concentrate. I looked out the window as my mind drifted, questions about my East German family entering my thoughts. Other than what we had learned from Albert's brief visit, and the sporadic, uninformative letters we received from the East, we knew little about the family.

Who were they? Who were they really? What were their lives like? How were they coping, living in a police state? Had they suffered any consequences from my mother's escape? Were they targeted, harassed for other reasons? My mother had a fighting spirit. Did they as well?

If they did, did their moxie ever find them on the wrong side of the regime? Did they ever confront the regime or did they survive by lying low and avoiding attention, toeing the Party line? I wondered if any of them were true communists or if they just played the game to survive.

Realizing that I would never know the answers to any of those questions, I shook off those thoughts, picked up my book again, and tried to focus on the incredible adventure upon which I was about to embark. But before long my mind wandered again and I put the guidebook back down in my lap. I just hoped they were all right and were courageous in the face of adversity.

I was, of course, strictly forbidden from trying to make contact with the family in the East. Any contact they might try to make with me would likely be a Stasi provocation or, if done on their own accord, could put their lives at risk. My mother's letters to them mentioned nothing about my career in the army, in intelligence, or even my assignment to Berlin.

I couldn't help but think about the irony of my situation: some three decades earlier, at the age of twenty, my mother had escaped the East, and now, at the age of twenty-two, I was going back deep into the very country from which she had fled. I was grateful to my mother for having taken the risk to run. What would my life have been like had I been born behind the Iron Curtain?

As we descended into West Berlin, I looked onto the fields and villages below. Suddenly the Berlin Wall came into view. Pockmarked with watchtowers, barbed wire, and armed guards patrolling the death strip, it was daunting and utterly foreboding.

I reported in to the Berlin Command and spent my first weekend learning the lay of the land. West Berlin was alive and bustling with activity. On the Kurfurstendamm, or Ku'Damm, West Berlin's main commercial boulevard, West Berliners enjoyed an amazing variety of restaurants, pubs and cafés, galleries, theaters, and parks. I went to the Kempinski for *Kuchen*, then to the zoo. Hundreds of people were doing their Saturday-morning shopping in the crisp autumn air. There were delicious smells everywhere of freshly brewed coffee, baked goods, whole chickens cooked on rotisseries, and *Würstchen* frying on outside grills.

Kaufhaus des Westens, KaDeWe, Berlin's "Department Store of the West," the largest department store in continental Europe, was a great display of opulence, a fantastic example of consumerism that showcased West Berlin affluence and free-market wealth. Its eight floors dazzled with the finest in crystal, glassware, and housewares; every level carried fully stocked shelves of cosmetics, perfumes, the latest Versace and Armani fashions. The top floor held a jawdropping selection of German and international foods, the choicest meats, seafood, cheeses, champagnes, wines, exotic fruits from all over the world, and the most exquisite desserts I had ever seen. I had yet to discover West Berlin's fifty square miles of lakes, forest, parks, and beaches.

West Berlin was a place to see and be seen. With its scintillating nightlife and parties, film festivals and cabaret shows, and eclectic avant-garde culture, it attracted people from all walks of life from all over the world, including famous movie stars and rock stars. Its economy clearly thriving, the city stood as a symbol of progress, freedom, and prosperity.

As I walked along the Ku'Damm, West Berliners and throngs of tourists bustled about under the colorful holiday decorations, enjoying life in a city they loved. Amid the signs and bright lights from the shopping district and theater marquees, West Berlin was a shining, modern city.

*T*he next day, I went to the Berlin Wall. The west side of the Wall was covered in graffiti. Urban artists, using the concrete façade as their canvas, had ignited a kaleidoscope of creative energy, of political and artistic expression. Alongside vividly colored, sometimes cartoonish images, spirited messages called for "Freedom!" or made simple commentary like "Why?" Other messages such as "Let them out!" berated East Germany as not so much a country as a prison. People from all over the world came to bond with others in a shared sense of disgust at what the Wall represented, or to simply splash color to offset the lifeless gray of oppression on the other side.

Along with other tourists, I climbed an observation platform and looked over the Wall into the East. The difference between the two sides of the city was jarring. Where West Berlin's heartbeat was strong, East Berlin's heartbeat had all but stopped. But for the presence of VoPo security patrols and very few cars, the city streets were murky and gray, the emptiness stretching as far into the distance as the eye could see, looking like an abandoned set from some old black-and-white 1930s film.

That fall, the Soviets shot down a South Korean commercial jetliner that had veered into Soviet airspace, killing all 269 on board. President Reagan referred to the brazen shoot-down as a massacre and as a crime against humanity. In speeches that year, Reagan began referring to the Soviet Union as an "evil empire," saying that communism "was the focus of all evil in the modern world." Reagan also unveiled the "Star Wars" missile defense initiative, aiming to develop new technologies to protect the United States from a Soviet nuclear attack. In response to the Soviet deployment of SS-20 intermediate-

Over the Berlin Wall: view of the lifeless East

range nuclear missiles capable of reaching Western Europe, the United States began to deploy Pershing intermediate-range missiles to West Germany, an upgraded nuclear missile system that worried the Soviets.

Both sides continued to introduce a steady stream of modern military capabilities in a dangerous game of military one-upmanship. The Soviet-made T-80 tank with its never-before-seen exploding reactive armor faced off with new U.S. M1 Abrams tanks and Bradley infantry vehicles on both sides of the Inner German border.

The largest, most concentrated armored force in the world now stood on East German soil, ready to strike NATO through the so-called Fulda Gap into West Germany. Five armies, some twenty divisions, air regiments, missile units, and nearly half a million troops,

all backed up by Warsaw Pact forces in Poland and Czechoslovakia, faced a much smaller NATO force that was outnumbered and
outgunned. With such a large and battle-ready force, NATO feared
Moscow could launch an attack against the West at any time. Forces
on both sides remained ever-vigilant.

In November 1983, NATO conducted a large-scale war game sim-
ulating a nuclear attack, which alarmed the Soviets. In Exercise Able
Archer, tens of thousands of U.S. and NATO troops participated
in the drill, which used realistic-looking dummy warheads and prac-
ticed realistic launch protocols. With East–West tensions already
on edge and, given the ongoing deployment of Pershing missiles to
Europe, Moscow became concerned that the exercise was a possible
actual prelude to war.

When NATO forces went to radio listening silence and the
Soviets could no longer track communications, Moscow became
alarmed that a preemptive attack by NATO was imminent. They
put their forces on high alert. Ten days after it began, the exercise
ended, leaving the Soviets greatly relieved, with NATO probably
not fully appreciating how close the two sides had come to war.
Many historians agree that it was perhaps the closest the world has
come to nuclear war.

Given the unpredictability of the Soviet threat, it was critical that
the Allies have a clear picture at all times of the enemy's intentions
and activities. East Germany, an area roughly the size of Ohio, was
teeming with hostile military activity that the United States and her
Allies needed to carefully track in order to guard against attack.

It was no wonder then that by late 1983, Berlin was a hotbed of
espionage and had become known as the "spy capital of the world."

Western intelligence agencies used Berlin as a main base for information collection, often resulting in a dangerous game of cat and mouse. Every intelligence collection method imaginable was used to monitor the Soviets, from listening in on Red Army and Soviet diplomatic communications to gleaning reports from a network of agents recruited to work on the other side of the Iron Curtain. It was an unforgiving environment where, in some cases, one small mistake or false move could easily put your or someone else's life in danger.

Deep inside East Germany, West Berlin was a perfect launching pad for intelligence operations into Soviet-controlled territory. U.S., British, and French national intelligence agencies used a vast array of collection methods to develop a clearer picture of Warsaw Pact intentions on their westernmost frontier. For a newly minted intelligence officer, it was the place to be.

Not to be outdone, Soviet and East German intelligence agencies—the KGB, GRU, and the HVA, the foreign intelligence arm of the Stasi—aggressively spied on the Allies in West Berlin and West Germany. Nothing was off-limits for them, including attempts to blackmail U.S. and NATO soldiers and diplomats, not unlike what they were doing with their own people, using sophisticated methods to learn about then exploit any perceived weaknesses, including problems at work, financial difficulties, or sexual weaknesses. Elaborate plots were launched to ensnare their victims and trap them inside interlocking rings of deceit.

On East Berlin's Normannenstrasse, like the grinding gears of massive machinery in constant motion, the main Stasi headquarters functioned at full bore. The gargantuan complex, made up of forty concrete buildings and labyrinths of halls and offices, housed a staff of more than thirty thousand Stasi officers, who worked in some forty departments, assigned round the clock to spying, reading people's mail, listening in on private telephone conversations, tracking targets, and recruiting agents.

Main Department VIII, known as Observation, kept an eye on East German citizens through its vast informant network, in which Stasi agents spent their time thinking up new ways to manipulate people and pressure them to spy on one another. Main Department II, Counterintelligence, carried out surveillance, including tracking and wiretapping of foreign diplomats. The Stasi even had a department to spy on its own personnel.

Throughout the East, the Stasi now employed some 90,000 people full-time, plus nearly 175,000 unofficial informants and some 1,500 others as moles in West German offices and in the West German government.

*S*everal miles away from Stasi headquarters, on the other side of the Wall, I worked at the American center of operations. The General Lucius D. Clay Headquarters Building, a German Luftwaffe headquarters during World War II, was named for the commander of U.S. Forces Europe after the German defeat. The sprawling three-winged complex now housed the U.S. State Department and the U.S. Berlin Command.

In my first job as the Berlin Command's intelligence briefing officer, I briefed scores of senior-ranking officials and politicians, including visiting congressmen and senators, about Berlin-based intelligence operations targeting the Soviets and East Germans.

One day, I briefed President Reagan's director of the CIA. Bill

Casey didn't appear especially intimidating as he sat at the head of the long, polished table in the conference room of the deputy chief of staff for intelligence. Casey was polite but sharp and asked a few pointed questions. At the conclusion of the briefing, Casey and his entourage left, and the major responsible for running a number of projects I had just briefed about approached me and asked if I wanted to swap my headquarters desk job for a chance to run real intelligence operations. I immediately accepted, but we both knew it would be an uphill battle as I would be the first woman to join the team. To my great relief, my nomination for the assignment was approved. I moved into my new office in the subterranean, windowless basement of the building.

The Operations Branch ran a number of collection programs, including ground and airborne intelligence missions. The aerial missions used a helicopter and a small fixed-wing aircraft.

Flying high, especially on a windy day, in a tiny, UV20 Pilatus turboprop airplane while focusing on the ground below through a thousand-millimeter lens, could be a challenge. The small, light aircraft, flown by seasoned U.S. Army pilots, bounced and lurched in the wind, causing the collectors in the back to fight against the movement and try to dismiss motion sickness in order to snap the perfect picture. Through the wide-open side door, harnessed to the aircraft, mission photographers leaned out of the airplane to photograph targets of interest in East Germany: industrial complexes, equipment storage sites, training areas, rail lines.

Our photos often gleaned valuable information, but the missions were not without risks. One day as we flew high above Berlin, our pilots carefully tracing the absolute edge of the airspace in which we were permitted by treaty to operate within, a Soviet An-2 biplane appeared almost out of nowhere and flew to within feet of us. As our pilot maintained control of our plane, we could see the Soviet pilot's face as he tried to intimidate us by repeatedly swerving his aircraft closer in, within clipping distance, then backing off in a crazy game of chicken. With orders to avoid confrontations, our pilots veered off and we returned to Tempelhof Airport.

Visiting East Germany was off-limits to all U.S. and Allied personnel unless on official orders. The Huebner-Malinin Agreement of 1947, however, allowed a small number of representatives from each of the four countries to venture into each other's sectors. Thus the U.S., British, and French had teams that regularly went into East Germany; reciprocally, the Soviets had teams that were permitted into West Berlin and West Germany.

The official mission of these teams was to exercise rights guaranteed under the treaty to travel and circulate. But everyone really knew the main purpose was being used by both sides to collect intelligence.

The Americans ran two ground reconnaissance programs that went into the East on a near-daily basis.

The U.S. Soviet Sector Flag Tours collected intelligence in the Soviet sector, East Berlin, and the U.S. Military Liaison Mission, better known as USMLM or the Mission, in the rest of East Germany. The Brits ran similar programs, BRIXMIS; the French, FMLM.

I was made team chief of the Soviet Sector Flag Tours. Because of the dangers of the mission, team members had always been men.

The Flag Tour team consisted of some fifteen U.S. soldiers, infantrymen, intelligence specialists, and professional drivers who were handpicked for their ability to be cool under pressure and have good judgment. The soldiers came from all walks of life and backgrounds—a tall Texas farm boy with a heavy southern drawl, a streetwise Latino from the Bronx, a bookish African American college graduate, a rough-and-tumble horse rancher from Georgia.

Critical to the mission was that they all work together as a close-knit team and have one another's backs at all times. Teams of two, a tour leader and driver, drove into the East in olive-drab American sedans. Although required by agreement to be identifiable by Allied license plates and a national flag decal, nevertheless the decal was small and the cars were subdued, designed to blend in and not attract attention.

Both Flag Tours and USMLM traveled in the East on city streets, on highways, and on rural roads and dirt trails as well—anywhere they could get a good picture or collect a valuable nugget of information. These missions were a vital opportunity for Washington to get information from up-close observations behind enemy lines. Our job was to have a good look around, to observe without being seen, to monitor and photograph anything that would help the Allies better understand the threat posed by the Soviets and East Germans.

Flag Tour missions crossed into the east through Checkpoint Charlie, the border crossing point made famous in John le Carré spy novels and scores of Cold War films.

Soviet and East German security agents were on top of us from the moment we crossed into the East. As missions gleaned valuable tidbits of information, constantly adding new pieces to the puzzle, they sometimes came with risks in the form of Stasi, VoPo, or KGB car chases, detentions, even deliberate rammings to force us to retreat or to scare us off. Soviet and East German soldiers were ordered to report Flag Tour sightings and encouraged to detain us, especially if we were found in sensitive areas, getting too close to something they didn't want us to see. We constantly weighed risk versus gain, but invariably it was hard to calculate the danger of any given situation and sometimes we would get caught or our team members would get injured. East German citizens were told to report U.S. Flag Tour sightings so that we could be tracked. We were spies, they were told, there to gather information that would ultimately be used to attack and destroy East Germany.

As Flag Tour team chief, I met up regularly with USMLM at their administrative headquarters in West Berlin to coordinate and plan operations into the East.

On one of my first trips to their offices, I met Major Arthur Nicholson. Nicholson was a boyish and academic-looking officer with a friendly demeanor and an easy smile. A veteran of more than one hundred tours, he was a rising star who had a reputation for being one of the army's top, specially trained Russian Foreign Area Officers. He was married and had a young daughter.

From the start, it was clear that Nicholson supported my selection as the first woman to serve in my job when some others did not agree with my appointment, and even resented the fact that a woman was assigned to the role of team chief of intelligence operations in the East. A consummate professional, kind and supportive, Nicholson was one of only a few who took time to mentor and professionally guide junior intelligence officers, including me. I was still new to my job when I first met Major Nicholson, who showed me around and introduced me to the mission team members. It was during that visit I learned that several months earlier French Mission Warrant Officer Adjudant-chef Philippe Mariotti had been killed in a car-ramming incident in East Germany.

Stasi photograph of Mariotti car-ramming

*O*ne hundred and fifty miles southwest of Berlin, in Karl Marx City, Paradise Bungalow was now producing an abundance of food, allowing Heidi and Reinhard to become independent from having to rely on the state for food. They relished their newfound freedom and turned a bumper crop that year of fresh vegetables and fruits, including zucchini, peas, and a variety of berries that were turned into ample quantities of jarred sauces and preserves. Soon their plum and cherry trees would be big enough to yield fruit, too.

More allotment gardeners erected sheds and cabins, so Reinhard helped to connect the patchwork of privatized bungalows to the nearby municipal power lines, giving them all access to electrical power for their radios and lights at night.

$\mathcal{B}$y now a teenager, Cordula continued to attend her special sports school and train on the national junior swim team. Like most young people her age, she craved good music. One day, she made a surprising and most welcome discovery.

She turned on the family's KR 450 radio to hear Honecker making an address from the annual National Youth Festival.

"You, the revolutionary guards," he exclaimed, "fighters for peace and socialism! Your unshakable will now more than ever is required." Uninterested, Cordula turned the dial, working it back and forth through the squelching until she picked up a faint trace of music. The reception poor, she came in closer, tuning the dial to hone the signal, and suddenly she had it. She had picked up West Germany's Bayern 3 broadcast from Munich, and the radio channel, it seemed, was airing something called the *International Hit Parade*. For the first time, Cordula heard Michael Jackson and Madonna, and West Germany's Nena singing "99 Luftballons." She listened to celebrity interviews and stories about sports teams in the West, teen social issues and relationships. The *Hit Parade*, she learned, was aired at the same time every week. It quickly became Cordula's favorite program and, as often as she could, she listened in.

$\mathcal{B}$ack in Berlin, one of my first missions opened my eyes to the dangers that came with the new job. In East Berlin, we had been directed to take a closer look at a rail line where we had been tipped off by USMLM to watch for trains coming through loaded with the latest high-tech Soviet air defense systems. We knew there had been a military training exercise in the area and were told to be on our toes.

The sun was setting when we made our way onto a winding path, which snaked into the woods and onto the back roads.

To get to the rail line undetected, we had to come in through a back way, through a patch of allotment gardens. The dirt road, meant for tiny East German cars and wheelbarrows, was extremely narrow and so the driver maneuvered our car cautiously, taking great care not to nudge a fence or drive into someone's cherished garden.

A woman kneeling, working in her garden, looked up as our sedan passed slowly by. When she noticed the tiny American flag decal on our car, she slowly stood. Instead of fear or aggression, she stood up tall, a look of solidarity washing over her face. As we moved slowly by, she looked at me and tipped her head in a barely noticeable but deliberate, measured nod, a show of support, the corner of her lips curling up into a faint smile.

Everything was going according to plan. All was quiet as we carefully proceeded onward. Suddenly the stillness was shattered by the thunderous revving of a large diesel engine at close range, the vehicle gunning in our direction. From a hiding place in the woods, some fifty feet away, a large, eight-ton, military-green TATRA truck lurched from its hiding place in the forest and came roaring toward us at breakneck speed, its power and large size easily capable of crushing our sedan.

My driver slammed the car in reverse and we sped backward in a spray of gravel and dirt. The TATRA lurched again, fighting to get to us, and just missed ramming our front passenger's side as it smashed through carefully groomed gardens, taking a chain-link fence with it. It was a close call. We were relieved as we made our way back onto the main road, then went on to check out a few other points of interest before returning to Checkpoint Charlie.

8 Spionagetätigkeit von Angehörigen der Militärinspektion (MI) der USA an Bahnanlagen innerhalb der Hauptstadt Berlin.

Spies on spies. These photographs of the author's team on an intelligence-collection mission in East Berlin were taken by the Stasi secret police. The caption reads: "Spy activity by members of U.S. military intelligence on train facilities in the capital of Berlin."

I told no one in my family about what I did in Berlin, except for my brother Albert, now a U.S. Army helicopter pilot (with a security clearance). When he visited me in Berlin I put him on a Flag Tour, which thankfully went off without incident.

Stasi photograph of USMLM on mission in East Germany

BRIXMIS on mission in East Germany

20

FACE-TO-FACE WITH HONECKER
MISSION IN LUDWIGSLUST
(1984–1985)

The situation in the world today is highly complex, very tense. I would even go so far as to say, it is explosive.
—*General Secretary of the Soviet Union Mikhail Gorbachev*

One day, Major Nicholson invited me and a few other junior intelligence officers to the USMLM house in Potsdam, East Germany. The Potsdam House was the jump-off point for USMLM teams setting out on tour in East Germany, and was also where mission personnel hosted receptions and ceremonies in their official capacity of furthering cooperation with their Soviet counterparts.

In the world of intelligence operations against the Soviets, USMLM tour officers and their British counterparts, BRIXMIS, were considered the best in the business. They were highly trained army, air force, and marine officers, usually majors, who spoke Russian and had academic backgrounds in Soviet studies, many having earned advanced degrees from the most prestigious universities. The French also had collection teams that were renowned for their daring.

I marked the upcoming visit on my calendar, and looked forward to seeing the Potsdam House for the first time.

By now, fifteen-year-old Cordula was swimming every day, training intensively among the country's best young junior athletes and hoping to one day get a chance to compete for a spot on the national team. East Germany entered every international competition with something to prove. In the 1980 Olympics, with the U.S. boycotting in protest over the Soviet invasion of Afghanistan, the East Germans came in a strong second place in both total medal and gold medal counts, just behind the Soviet team. In 1984, the Soviet Union, along with other Eastern Bloc countries, including East Germany, shot back by not participating in the Los Angeles Olympic Games due to what they called "anti-Soviet hysteria being whipped up in the United States." Despite missing the 1984 Summer Games, East Germany's total Olympic medal count between 1968 and 1988 would rival that of the Soviet Union and United States, and also outnumber West Germany's count by three to one.

With East German sports soaring to new heights, the world held its breath every time East German athletes competed; everyone expected something spectacular to happen and records to break. The immensely proud citizens of East Germany ecstatically cheered their country's phenomenal athletic successes, becoming temporarily distracted from the limitations of their lives and the hardships plaguing their country.

But behind the scenes, what Honecker and the country's cadre of trainers knew and the rest of the world, East German citizens, and even the athletes themselves often did not was that, while the regime and the country reveled in its newfound superstar status, all along the East Germans had been cheating. This was the world's first example of state-sponsored doping. While some of their athletes' successes could unquestionably be attributed to pure talent, radical cutting-edge approaches to stretching the limits of the human body, and super-intensive training practices, some of their top athletes were also gaining an unfair advantage by being fed performance-enhancing drugs.

As this information came to light, a pall fell over all East Germany's athletic accomplishments, and a spotlight of scrutiny cast doubt every time an East German shattered a world record. Now the police state with the dismal human rights record added another failure to its list and saw its global reputation plummet.

*O*n October 7, once again the haunting gray hues of the country gave way to an explosion of color to show off the military might of the communist state, this time to celebrate the thirty-fifth anniversary of the founding of East Germany. As usual, Honecker trumped up the country's progress, proclaiming, "This country is nowadays one of the most advanced industrial nations in the world."

Roland, Tiele, and all my mother's other siblings and their families attended the annual parade in the cities and towns where they lived. Heidi, Reinhard, Cordula, and Mari, now a Young Pioneer, as usual, attended in Karl Marx City.

In East Berlin, I was one of a handful of official U.S. representatives authorized to attend the parade. On a sunny, chilly morning, tens of thousands of people packed Karl Marx Allee in East Berlin, having been bused in from districts in and around the city. The crowds were tightly packed, spanning five rows deep. With plainclothes Stasi and police militia integrated into the assembly of people and milling about to keep an eye on the spectators, a sea of paper flags waved: sky-blue ones with white doves; flaming red ones with hammer, sickle, and star; and the national flag, the tricolor red, black, and gold with hammer and compass, encircled by a wreath of wheat. Up above, red banners proclaimed, "*Starker Sozialismus, Sicherer Frieden,*" "Stronger socialism, more secure peace."

Off to the side, a whole section of archetypal communist youth, fresh-faced children, dressed in bright, spotless clothing and red and blue bandannas, waved, cheerleaders for the assembled masses. Up above in the reviewing stand stood the master of ceremonies, the country's leader, Erich Honecker. Soviet foreign minister Andrei Gromyko and rows of Party officials acknowledged the crowds with nods and waves as people cheered loudly in a forced show of enthusiasm on mostly blank faces. Beyond the reviewing platform of important dignitaries, concrete-block-styled buildings in various stages of disrepair lined the street in odd juxtaposition to the exaggerated display of confidence and capability playing out on the parade route below.

A general festooned with rows of medals offered an official welcome in a tinny address that reverberated through the cranked-up sound system. While Party officials gave propaganda-laden speeches, criers planted in the crowd told the assembled when to cheer, when to applaud and wave their little flags, while police and undercover security monitored the scene, keeping an eye out for troublemakers. To my right stood a group of teenage boys looking bored. When one noticed they were being watched by a plainclothes monitor, the boy nodded and suddenly feigned interest, waving his flag back and forth, his friends following his lead.

Finally the parade kicked off. Drummers, bands, flag-bearers, NVA troops squared off in blocked groups of gray uniforms, wearing white gloves and flat gray helmets, rifles clasped across their chest, chins raised in resolute loyalty, made their way up the parade route, jackboots precision-stepping to the beat of blaring nationalistic music. Rows of soldiers marched past, followed by the most powerful weapons in the East German arsenal. Soviet-made self-propelled howitzers and T-72 tanks as MI-24 gunships flew in an impressive tight formation over our heads. Few seemed to notice me as I stood inconspicuously among them. Instead, people focused on what they were supposed to be doing: paying attention, cheering, and waving their little flags.

Along with the crowd, I watched as the equipment rolled by. I had already seen some of it close-up. Three hours earlier, as the vehicles were being staged farther down the road, our team had been taking pictures. We had carefully jostled around young East German soldiers who had tried to prevent us from getting close to the equipment. At one point, my team members feigned sudden interest in something in the distance and excitedly hurried off, distracting the East German soldiers who hurried off in pursuit of them. Their stunt worked and I was left unguarded to take close-up pictures of a new model of a Soviet-made, tracked, infantry fighting vehicle.

At the end of the parade, after the last vehicle had passed and the crowd began to thin out, I wandered around a bit and came upon a gathering of important-looking dignitaries. I approached to investigate. About a hundred stately looking officials in dark suits—Party elite, I assumed—stood clustered near a war memorial. They stood in stoic silence and appeared to be waiting for something or someone.

ZSU 23-4 antiaircraft gun, East Berlin

I casually walked over to the group, bypassing security guards, and slid in to take up a place among them, as they inched back up into one another to make room for me, apparently believing that I was meant to be there.

After a few minutes, a fleet of shiny, black Russian-made limousines shot up to the curb, stopping with great precision just inches away from one another. Men in black scrambled about, hurriedly opening doors for gray-haired dignitaries. Among them was a wiry, bespectacled man, his white hair neatly combed, glasses set securely on his temples, framing a pale, solemn face.

I recognized him immediately. It was Erich Honecker and he was standing ten feet away from me. Tightly surrounded by a keenly attentive security detail, Honecker, his deputy Egon Krenz, and others moved swiftly up the path as one unit, bypassing me and the rest

Stasi and East German security keeping a close eye on the author in East Berlin

of the crowd, and made their way to the memorial. No one moved; there was no applause, no sound, only silence.

At the memorial, Honecker placed a wreath and ceremoniously bowed his head in silent tribute. Seconds later, he and his entourage began the walk back to their waiting motorcade. As Honecker walked back up the path, when he was just a few feet away, I half-stepped out from the crowd. For a split second, the leader of East Germany and I made eye contact, and I snapped a picture.

It was an epic moment for me. It lasted just seconds. I was there. He passed by. He looked at me. And then it was over.

Back at the office, the photo revealed the contemplative gaze of a man lost in thought.

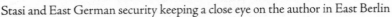

To his people, Honecker appeared confident and in control. In truth, however, he was worried. By now the Solidarity labor movement in Poland was building momentum and the East German leadership feared the growing protests there might spread to East Germany. To make matters worse for Honecker, the Soviet Union had a brand-new, youthful reform-minded leader. Mikhail Gorbachev, acknowledging that his country was in decline, corruption was rampant, and the country was being smothered from the weight of it own inefficient communist bureaucracy, decided to implement a series of political, economic, and social reforms in the Soviet Union.

East Germans started to take notice when they heard Gorbachev's speeches. With great interest, they followed his ideas about change in the Soviet Union, hoping to one day see similar reforms in their own country as well. But Honecker had no intention of following the Soviet leader's example. He rejected Gorbachev's ideas, opposing any change that might put an end to his totalitarian regime. But now he would have to find a way to address a puzzled public who had been conditioned to view the Soviet Union as a role model and a big brother to emulate. The East German leadership could hardly criticize the Soviet Union, but neither could they encourage their citizens to follow Gorbachev's model. So Honecker clarified his position: following the Soviets, he said, no longer meant following their *every* move. With that sentiment, the East German government dug in its heels, determined to hold out.

*I*n West Berlin, I and a few other officers met up with Major Nicholson, who took us to the historic USMLM house in Potsdam. From West Berlin, we made our way into East Germany over the strictly controlled Glienicke Bridge.

As we drove over it, the lone car on the vacant stretch of that storied landmark, Major Nicholson talked about the role the bridge played during the Cold War. Half in the East, half in the West, the Glienicke Bridge was known as the "Bridge of Spies" for being a place of choice for major Cold War spy swaps and prisoner exchanges, including the famous 1962 exchange of U.S. pilot Francis Gary Powers for Soviet spy Colonel Rudolf Abel.

At the Potsdam House, we toured the carefully groomed grounds, and inside the residence saw the mansion's spacious rooms, where the mission hosted diplomatic receptions for U.S. and Soviet dignitaries. We toured the house and ate lunch before crossing back over the Glienicke Bridge and returning to our offices in West Berlin.

Less than a hundred miles away, in quiet, remote Klein Apenburg, Opa's health had declined. At eighty-six years old, he was weakened from old age and diabetes. Roland took him to the Apenburg hospital, where, as a result of complications from the disease, he had surgery to amputate his leg. On the operating table, my grandfather passed away.

Now the patriarch of the family was gone. Opa's life had spanned most of the twentieth century in a period that had seen incredible change. He had been a soldier in both world wars, had tried to adapt first to Nazism—privately he called Hitler "a madman who despised human beings"—and then to communism, only to ultimately end up fighting the system. The family laid him to rest beside Oma in a cemetery near Apenburg. Roland announced Opa's passing to Hanna in a letter.

Cordula took the loss of her grandfather very hard. She attended the funeral but had to report back immediately for training. Women's cycling had become an Olympic sport in 1984 and her trainers

Major Arthur Nicholson with Soviet counterpart at
BRIXMIS reception

had just given her the opportunity to switch her sport from swimming to cycling. Though Cordula had had little experience other than the riding she had done on the family's old single-gear utility bicycle, she disciplined herself to focus and showed remarkably fast progress. She worked hard to master the East German Diamant road race bike and before long was competing in races throughout East Germany.

In the Soviet Union, General Secretary Gorbachev emerged as a new kind of leader. President Reagan recognized in Gorbachev a different mind-set and the potential to help forge a new era in U.S.-Soviet relations, one of genuine progress between the two major world powers. Though there were still fundamental chasms between the two superpowers, for the most part the relationship took on a more civil tone and for the first time there was a real sense that change might indeed be possible.

Then something unexpected happened in East Germany, which once again spiked tension in Soviet-U.S. relations.

One morning in late March 1985, I came in to work to prepare for the day's operations into East Berlin and was informed that there had been an incident involving a USMLM tour officer who was on a mission in East Germany. Major Nicholson was dead.

21

BEYOND THE CHECKPOINT
PASSAGE
(1985)

> We can only hope that the Soviet Union
> understands that this sort of brutal international
> behavior jeopardizes directly the improvement in
> relations which they profess to seek.
> —*Vice President George H. W. Bush, upon the return
> of Arthur Nicholson's body to the United States*

*M*ajor Nicholson had been shot by a sentry near a Soviet military training area while on a mission near Ludwigslust, in northern East Germany. With his driver providing eyes and ears lookout, Nicholson had stepped out of his vehicle and approached storage sheds used to house tanks. Neither Nicholson nor his driver had been aware that they were being observed by a Soviet soldier, a young conscript armed with an AK-74 rifle who followed the two in his sights from his hidden position in the wood line.

Without warning, the young soldier aimed his rifle and pulled the trigger. The first bullet whizzed over the heads of the tour. The second struck Nicholson in the abdomen, dropping him to the ground. When the driver jumped out of the car with his medical bag, the sentry closed in, rifle pointed at the driver's head, screaming for him to get back in the car. While the sentry held the driver at bay, a dozen Soviet officers and soldiers arrived on the scene, but no one made a move to approach Nicholson, who lay on the ground bleeding to death. More than one hour after the shooting, a medic finally appeared, crouched down to Nicholson, took his pulse, and said, simply, "Nyet."

Major Nicholson's body was delivered back into U.S. hands over the Glienicke Bridge.

*F*atal Shots Without Warning," "Casualty of the Cold War," "Soviets Offer Apology in Killing of U.S. Major." The incident made headlines around the world and brought the progress of Soviet-U.S. relations to a screeching halt as Secretary Gorbachev faced his first crisis as leader of the Soviet Union. The Reagan administration confronted Moscow, warning that the episode jeopardized the improving relations between the two countries. Before long, however, the incident was downplayed as an unfortunate event that should not disrupt such a critical period in superpower relations, at a time when both sides anticipated the first real strides toward an end to the Cold War.

*I*n the Soviet Union, Gorbachev began to institute his plans for restructuring (*perestroika*) and openness (*glasnost*).

As Gorbachev explored change at home, he began to reach out to the United States and other Western nations. The United States welcomed the new Soviet leader's desire for change. The world watched as a new optimism took hold and relations between the two super-powers showed signs of improvement.

True to his plan, Gorbachev began to institute change in the Soviet Union, urging other Eastern Bloc countries to follow his lead. The Soviets now admitted to a crumbling economy, decaying infra-structure, poor housing, food shortages, alcoholism, and increasing mortality rates, as well as the country's infamous history of state crimes against the population. Topics like the legacy of the KGB, once taboo, were being openly debated.

Honecker, meanwhile, admitted to none of his country's short-comings. He remained unmoved, refusing to alter the status quo in East Germany, publicly saying, "We have done our perestroika. We have nothing to restructure."

Despite appearing unmoved in public, Honecker felt betrayed by Gorbachev. Honecker eluded talk of reform and instead started to censor Soviet media in his country, even banning the Soviet mag-azine *Sputnik* and Soviet films, and ordering that new school text-books avoid the topic of the transitions under way in Russia.

Obdurate and dogged, Honecker remained rooted to his cause, vowing if necessary to be the last remaining leader of hard-line com-munism in Eastern Europe.

By the time she turned sixteen, Cordula's cycling career had taken off, and she entered the ranks of the East German national team. The newest and youngest member, she trained three times a day, six days a week, alongside her more experienced teammates, all tried and tested Olympic athletes and hopefuls whom Cordula worked hard to emulate.

Her days were spent sequestered in intensive training, her schedule completely regimented. She did everything with her new teammates:

East German teenager Eike Christine Radewahn attempted to swim to
freedom across the Danube River in Romania. Border guards shot at and
then arrested her. She was sentenced to three years' imprisonment
in East Germany's Hoheneck Castle Prison.

training, eating, sleeping, and back again to training. Wind sprints,
weight lifting, gymnastics, and constant track, stationary bike, and
road work commanded her days. Bedtime was nine o'clock, lights out
at ten sharp, wake-up well before dawn. The sports program invested
completely in its athletes, sparing no effort and overlooking no detail
to monitor every facet of the girls' lives in an attempt to propel success. Outside influences, including television and film, were screened,
and not allowed if they distracted and did not serve to inspire and
motivate. Aside from the occasional sweets, the girls were not allowed to consume anything that would interfere with shaping an ex-
traordinarily superb sports psyche and optimal physique.

For Cordula it was a promising beginning. Over the next months she
made remarkable progress, her diligence and determination rewarded
when she was chosen to ride in her first two competitions in Poland and

Czechoslovakia. To the great satisfaction of her coaches, she continued to win medals at local and national events throughout Eastern Europe.

Then one day, the family got some utterly stunning news. Heidi took the call Cordula made from the athletic training facility in Leipzig.

"Mutti," she said, "I made it. I made the team."

Numb, Heidi slowly sat down. East Germany had bestowed one of its highest honors upon Cordula. She had landed a coveted spot on the Olympic training team. Catapulted to the top level of the elite world of East German sports, she would now compete in world championships and contend for one of three spots on the Women's Olympic cycling team.

Membership in the elite East German sports system would mean Cordula would have to buy in not only to the rigors of training and the pressures of performance, but to the ideals of the regime. On the other hand, thought Heidi, it would mean that Cordula would likely get a chance to travel to the West. Just maybe, Heidi thought, her daughter would be able to see what Heidi herself had once seen when she was five: what life was really like on the outside. There was great pride in the entire extended family when they learned that one of their own, their daughter, sister, niece, cousin, Cordula, had earned a rare opportunity, one that most could only dream of, and they hailed it as a huge victory for the entire family.

At the track, Andrea, one of the stars of the team and an unofficial leader, and the rest of the members of the women's cycling team congratulated Cordula with a businesslike welcome that underscored the seriousness with which they took their roles. It was a day Cordula would never forget. Her country believed she had the potential to

be a world-class athlete, that she was among the best her nation had
to offer, that she could bring home gold for East Germany. By East
German standards, she had made it to the top.

A few weeks later, after much deliberation, Heidi finally made
a painful decision. Not wanting to jeopardize Cordula's once-in-a-
lifetime opportunity, she severed all communications, her once every
year or two letter exchange with the big sister she hardly knew but
had come to adore.

Sitting in her little apartment in Karl Marx City, Heidi penned a
final letter to Hanna, telling her only what she could, knowing the
Stasi might examine it carefully for clever wording and surreptitious
double meanings.

"Cordula has been given a chance at a good life," she wrote, "so
this will be our last contact. I know you will understand." If the authorities read the letter, they were no doubt satisfied, but Heidi was
heartbroken.

After cutting ties with the sister she had met only once but who
had become a source of strength, her role model for courage and independence, Heidi consoled herself by knowing she had done the right
thing for Cordula. Heidi asked her siblings to keep her informed of
any news from their sister.

In Washington, my mother received the letter and had no idea
what it meant but knew that, for Heidi to cut off all contact, something big had to have happened.

*I*n the national sports training facility in East Berlin, athletes assembled from all over the country, from Leipzig, Erfut, Cottbus, and
Karl Marx City. A ministry-level sports director took to the stage

and the room fell silent. Holding a microphone, he turned to address the room packed with the latest fresh crop of the nation's best athletes. His voice weighted, his speech measured, he delivered his words with great calculated effect:

> You are no longer common citizens of East Germany. You are now the nation's elite.
>
> You have been selected to represent the top tier of our society. Your performances in your sport will bring great rewards to you, your family, and to your country.
>
> With this honor, comes great responsibility. Each of you has been chosen for your athletic ability to be not only the best in East Germany, but to be the best in the world.
>
> Your job is to work harder than everyone else. Let no one outdo you. Commit yourself completely and without reservation every day toward every task. You are the pride of our country. Make your trainers proud. Make your country proud.

Athletes training for the Olympics had special benefits. Besides travel, they enjoyed foreign goods, things that were unavailable to average East German citizens, like the latest electronic gadgets and household appliances, Western shoes and clothing, cars, and so-called specialty foods. While average citizens stood in line for a limited selection or patchy supply of food, the athletes enjoyed nutrition necessary to build strength and muscle mass, their diets included

Western imports, such as bananas, oranges, and a variety of energy-fueling carbohydrates and strength-inducing proteins.

They also had access to world-class athletic equipment. For Cordula that meant riding superbly designed bicycles with hand-crafted Italian frames with top-of-the line Japanese components. Sports and cycling apparel were engineered with the most advanced aerodynamic synthetic fibers of the day: polyurethane, Lycra, and polyester blends, including lightweight, high-performance fabrics. Footwear, cycling, and running shoes were West Germany's Adidas World brand, the leading-edge sports shoe of the day.

Athletes trained in optimal conditions, had strictly controlled diets, and were constantly tended to by the country's top sports physicians and physiotherapists, who were always on site to treat, mend, and see to their every need. Some twelve thousand coaches served East German athletes, compared to fifteen hundred that managed athletes in West Germany—nearly ten times as many. Hundreds of psychologists were responsible for individual and team dynamics and for making sure the athletes toed a strict ideological line.

For Cordula and all the athletes, it was a whirlwind of excitement. Although the perks were amazing, more than anything she felt proud to be among the best athletes in one of the most rigorous, hardworking, and successful sports programs in the world.

Fully incorporating their cutting-edge sports methodologies, in every training session trainers drove their athletes to the absolute limit, demanding they give their best every single day, and the athletes themselves felt compelled to excel, always aware that anything less than consistently superior performance would get them tossed out of the program and replaced at the snap of a finger by someone who could reach higher. Constantly pressured to achieve superb results

Cordula prepares for a race.

and with the regime relentlessly driving them, the athletes could see propaganda-laden billboards and banners hanging from almost every surface of every Olympic training gym, urging them on and reminding them at every turn, WE ARE WORLD CHAMPIONS! and WIN! WIN! WIN!!!

One day in the fall of 1985, Cordula, the Olympic training team, and their coaching staff came to East Berlin to train at a velodrome not far from where I was conducting intelligence operations.

After a good night's rest in their dormitory near the Marzahn Sports Complex in East Berlin, they had breakfast and began the day with a surprise. Before practice, the trainers would treat the girls to a VIP tour of East Berlin.

As usual, the streets of East Berlin were eerily desolate, devoid of pedestrian and vehicular traffic, aside from the occasional policeman or roving patrol car. Wearing their stylish azure-blue team jackets emblazoned with the white letters *DDR* (East Germany) and country patch with the state crest, the privileged athletes, flanked by their trainers and a couple of security men, walked up the famous Unter den Linden. Once the center of downtown Berlin, the storied chaussée had been a thriving cultural center, an elegant boulevard lined with blossoming, fruit-bearing linden trees and teeming with life as people strolled about on the green or passed lively cabaret houses, embassies, and crowded cafés on their way to the opera or museums. But, like the rest of East Berlin, Unter den Linden had fallen into disrepair.

Despite the recent construction of high-rise apartment complexes and nominal restoration of some of the older buildings, the boulevard remained a bleak street with drab façades and tenantless buildings, the entire area empty of the life that had once flourished there. Cordula and the group moved westward, down the avenue. Before she came to the end of the road, she saw it.

Rising out of a haze in the distance stood the colossal Brandenburg Gate, the iconic symbol of German strength and unity. Originally built as a gateway through the heart of Berlin, and the city entrance leading to the palace of the Prussian monarchs, the towering hundred-foot-tall golden sandstone monument, with its twelve Doric columns and patinaed *Quadriga*—a statue of the goddess of Victory commanding a chariot of horses—now stood in the East, all access through the colonnade blocked by a thick, massive concrete structure, the Berlin Wall. Beyond the Gate and the Wall lay the city of West Berlin.

Fifty yards from the monument, the trainers halted their athletes at the rope that cordoned off further access. Being so close to the border, the site was strictly off-limits to the East German public. On this day the area was deserted, except for the presence of the vaunted athletes, their minders, and a few armed guards who lingered nearby, prepared to act should anyone attempt escape. Standing in the gray morning mist, the athletes looked on, Cordula seeing the border for the first time.

They stood in silence, contemplating the view, gazing along the spotless expanse of concrete to the Gate, the Wall, and on into the West, while the trainers stood quietly by, stealing glances at the girls, trying to detect even the slightest expression of want in their faces.

One of the trainers broke the silence, reminding the girls how fortunate they were to live in a country that had built a barrier to protect its people from the evil and depravations that lurked beyond it. As he spoke, Cordula nervously eyed West Berliners and tourists on an observation platform peering over the Wall into the East, and she was relieved to be at a comfortable distance from the frightening West. With international competition only weeks away, the trainer uttered the usual bromides about defection, deriding the "poor, misguided souls" who had tried to escape to the West. He added the usual warning: "It can only result in tragedy for those who attempt it and for those they leave behind." He left them with one last message, one the athletes had long since learned by heart: "Leaving the East is a betrayal. It is the ultimate betrayal to yourself, to your family, to your teammates, to your trainers, and especially to your country, which has placed its complete trust in you and has given you everything."

By now the trainers had good reason to be concerned, as hundreds of athletes had already defected.

After seeing the sites, Cordula and the team moved to the Werner Seelenbinder velodrome, where they trained in the gym and on the track over the next couple of days. At around the same time, in West Berlin, our two-man Soviet Sector Flag Tour team headed out on another intelligence collection mission in East Berlin.

The morning fog dissipated, replaced by a cool, overcast day. Less than a mile from the Brandenburg Gate, and four miles west of the velodrome, my driver and I approached Checkpoint Charlie in our sedan. He picked up the mike and called in a communications check with our base in West Berlin. I checked one last time to ensure that our equipment, cameras, binoculars, maps, and notes were all out of sight, windows up, doors locked, then zipped up my jacket, pulling the collar up closer around my face. I acknowledged my driver, he nodded back, and we moved through.

As usual Checkpoint Charlie was quiet, desolate, and charged with a chilly, foreboding tension. All color seemed to fade as one approached the infamous checkpoint, the world's foremost perilous front line in the Cold War and the absolute edge of where the free West and communist East stood head-to-head in an ideological showdown. Preparing to enter hostile territory, we always faced the reality that any mission could take an unexpected turn, putting us smack in the middle of a hostile detention, a dangerous car ramming, or worse. With no lifeline other than occasional, spotty radio connection with our headquarters, our teams went into the East alone, fully understanding the risks and knowing that there could be little chance of rescue should something go wrong.

Approaching from the West side of the checkpoint, we moved forward slowly, bypassing armed American Military Police, and continued forward. Besides the MPs, only one other person stood on the West side of the checkpoint, a gaunt, young German woman holding a placard that read: "Hunger strike until they return my children to me."

We drove onward, passing staggered concrete barriers and barbed wire and the sign written in four languages, warning "You are leaving the American Sector," cautioning those going any farther that they would no longer be protected by the safety of the West.

As we made our way through the barren corridor, high above our heads, up in the watchtower, East German sharpshooters scanning the Wall for escapees in the death strip below now trained their attention on us. Through binoculars, they tracked us, following our careful, measured movement through as we reached East German border guards. At the East checkpoint, a stone-faced guard picked up the phone receiver and made a call, alerting surveillance of our crossing. Then he raised the metal gate barrier and we slipped onto East German territory.

As usual, surveillance picked us up immediately after we crossed over. Sometimes it was VoPos, sometimes other security or intelligence services driving Ladas, Volgas, or Trabants, but this time it was Stasi, a two-man team in a small, pale green unmarked East German Wartburg, that fell in neatly behind us and followed at a close, tailgating distance. We turned onto a main street, Leipzigerstrasse. Before us, East Berlin streets lay open, as usual devoid of traffic. We accelerated, then turned onto a side street. They stayed with us, but we sped up. Then we turned again and stepped on the gas, the team on our tail working hard to keep up, which they did for a while, until we outran them a couple of miles down the road, their tiny underpowered, two-stroke engine no match for our powerful Ford.

Up the road, as we expected, another tail waited on a side street, ready to pick up where the last left off. It sputtered into action, flying out from its hiding place, and raced behind us for a while, its gears grinding as it tried to keep up with our pace. The car followed for a stretch, but a mile or so in, it broke down on the side of the road, smoke pouring from its engine.

Once in the clear, we drove along for a few miles through the center of East Berlin, breathing in the heavy air, thick with the acrid smell of burning coal, as we made our way on a vacant stretch of road that, despite being in the middle of a city, saw few other cars and only an occasional human form. Against a sooty, gunmetal sky, autumn always made East Berlin appear even more sad and lifeless. Roads, sidewalks, and storefronts were all but deserted, buildings decrepit, some boarded up, windows sealed off with concrete blocks, rooftops laced with barbed wire, some with spiky shards of glass.

Farther into the city, mass-produced Soviet-styled utilitarian high-rise apartment blocks were presumably filled with East Germans, but strangely, few people could actually be seen. Befitting a police state, East Berlin, the capital of East Germany, was a colorless city made up of harsh hues of black and gray, a microcosm of an entire nation that was hiding under a strange veil of secrecy.

At the velodrome, Cordula and the rest of the East German women's cycling team prepared for an intensive training session. The coaches began the day by putting the girls through a series of short exercises. Their single fixed-gear bikes having been carefully calibrated, tuned, and inspected by a cadre of technicians, the athletes, dressed in skintight Lycra, took up their positions on the track. Sit-

East German city landscape: high-rise concrete apartment blocks with
Honecker poster in the foreground

ting astride their bikes, balancing and steadying themselves between
their trainers and the railing, they readied for the launch, the first of
a dozen runs they would make that day.

With stopwatch in hand, one of the trainers blew a whistle and
the girls took off, accelerating as they made their way around the
250-meter pinewood track, their thin tires affixed to the track's steep,
slippery surface by centrifugal force. Winding their way around the
track, they picked up speed as the trainers shouted, demanding
better technique and maximum effort. The girls worked harder, increasing speed on the straight, angling to whip around the curve. At
the second pass they began posturing for position, and by the beginning of the next lap, their trainers were shouting, "Weiter! Schneller!"
egging them on to go even faster. The girls responded with each push

of the pedal, jostling for control of the track and finally exploding in a burst to finish the heat.

As the track portion of their training day neared its end, all six team members were pitted against one another in a series of sprints designed to breed internal competition and keep the top riders on their toes.

$\mathscr{A}$ few miles away, my driver and I reached the district of Karlshorst, home to East Berlin's Soviet military headquarters and to the Soviet Sixth Independent Motor Rifle Brigade. Believing we had not been followed, we continued on to our target.

We left the main highway and took a narrow trail through a thick forest, a trail that our team had used once before that had been carefully chosen during hours of mission planning to help conceal our movement so we could reach our target unobserved. We moved deeper into the forest, trying to avoid ruts in the road and winding our way along the bumpy dirt path never intended for use by a passenger car, all the while scanning the wood line for signs of danger.

Everything seemed to be going according to plan as we approached to only a few hundred feet from our objective.

Just as we moved into position from where we were to observe activity taking place near the headquarters, a single Soviet soldier, weapon raised, stepped directly into the path of our oncoming car only giving us enough space to hit the brake and come to a halt. Other soldiers appeared, taking up positions around the car, surrounding us and cutting off any chance of escape. Now a fair distance inside the forest and well outside of radio range from our headquarters in West Berlin, we were in a bit of trouble.

With the soldier in front still blocking our path, another strutted over to my side of the car, holding a pistol as he chambered a round. I slowly looked up to see that he was a lieutenant, no older than me, and like me, he was in charge. Pumped up with an air of exaggerated self-importance, he nodded, taking his time to relish the moment, his catch, an achievement that would likely win him praise from his commander and his fellow soldiers. A condescending smirk spread across his face.

I returned to looking straight ahead and sat motionless. The next thing I knew the muzzle of his loaded pistol tap-tap-tapped against the glass only inches away from my head. He ordered, "*Atkroy okno.*" (Roll down the window.)

I slowly held up my hands in the air in a gesture of defeat, but did not do as he commanded.

"*Seychass!*" (Now!) he shouted, the muzzle of the gun fixed against the glass and pointed at my head.

I showed no reaction to his aggression, but my mind raced to process the situation. Sitting deep inside an East German forest, we had few options, but I knew that the longer we sat, detained at gunpoint by this band of young, excitable Soviet soldiers, the greater the likelihood that things would spiral out of control.

I slowly panned the scene, trying to take in any details I could turn into a possible advantage. I noticed that they did not appear to have a radio, so I slowly moved my hand toward our Motorola radio handset, careful not to provoke, each inch of the way sensing his reaction, as he stood there, as they all stood there, guns at the ready, staring, riveted to my every move.

Though we were by now out of transmission range, I picked up the mike and, ever so cautiously, feigned a call back to our base in West Berlin, gambling that the lieutenant might think twice about making any rash move if he believed I had reported our situation and location.

"Base, this is one-o-four. I have a situation to report. We are being detained in the vicinity of Karlshorst by armed Soviet soldiers."

*B*ack at the Marzahn velodrome, Cordula and her teammates prepared for the next round of races. On the track, the athletes once again mounted their bikes. The clock started, the whistle sounded, the trainers released, and the cyclists took off. This run a controlled exercise to sync technique with speed, they worked their first laps at a measured pace. Increasing their rotations, they moved faster, picking up more speed. By the fourth lap they were full-on, pushing with everything they had, their tires defying gravity on the track's sloped walls, as they blocked and postured for advantage. Then, just when they believed they were at maximum exertion, one of the trainers pushed the button on a cassette tape player hooked up to a loudspeaker system and cranked up the volume, the loud music signaling them to take it up another notch. The trainer continued to increase the volume until the speakers could no longer take it and the sound became distorted. It triggered a reaction, the one the trainers were looking for, the cyclists' speed and thrust now forced to an impossible level as they moved faster and faster, forging full-on with everything they had as they rode to the earsplitting lyrics of Queen's "We will, we will rock you," the *bang-bang-boom, bang-bang-boom* of the bass fueling them. Pushing, stretching, pedaling as fast as they could go, fighting as they gapped, dropping back an inch, they surged forward, trying desperately to control their bikes, finally exploding in one last burst of speed until the trainers clocked it out with Andrea, as usual, riding cleanly to the finish and winning the race, the other girls streaming past, one after the other in her draft.

*I*n Karlshorst, everything was at a standstill. There was only silence, jarring and intense, like the stun lock moment before your parachute deploys. With a gun pointed at my head through the car window, I was alone with my thoughts, flicking through my mental files and relying on my instincts to try to gauge the soldier's state of mind and his maturity as a leader. During and in the aftermath of the Nicholson incident, the Soviets had reacted with blatant disregard for the loss of life, so it was hard to predict what he might do, but I kept trying to sense whether he was likely to pull the trigger. Did he have any respect for me as an adversary, as an American, as an officer, as a woman, as another human being?

I held the mike, trying to bide my time, pretending to talk and listen for instructions from home base. I knew I needed to somehow get control of the situation, to get the lieutenant to calm down before he or his men did something reckless. I sensed that he was waiting me out. As I "consulted with base," I slowly turned to look up at him and into his face.

The gun, still pointed at my head, now began to shake, and rattled against the glass as if he had become nervous or confused. I seized the moment, turned back slowly to my fake radio transmission, talked and nodded, then gently but decisively set the mike back into its clip holder. Acknowledging with a slight nod that the lieutenant had the upper hand, I then made a slow but firm motion, as deliberate a hand gesture as I have ever made, pointing to show him my intention to move the car forward.

"Inch the car forward," I told my driver.

"Ma'am, there's a Russian with an AK standing in front of us."

"Inch the car forward," I repeated. "He'll move."

My driver carefully pressed the gas pedal and the car gently pushed up against the soldier with the rifle who was blocking our escape. His eyes grew wide. He reacted by tightening his grip on his weapon and thrusting forward, butting up resolutely against the front of the car. He looked over at the lieutenant for guidance.

"Keep moving," I ordered, "very slowly." As we continued moving forward, there was a nervous exchange of words between the soldier in front, now being forced by our car to walk backward, and the lieutenant, who had started walking to keep pace with us, trying to keep his pistol aimed at my head. Suddenly he stopped walking and called out an order and the soldier in front stepped away from our path with a disappointed look on his face, as if a fox had slipped from the sights of his hunting rifle.

22

IMAGINE
THE ROAD AHEAD
(1986)

History has proved that Lenin was correct.
—*East German tenth-grade civics book*

The gun fired and we were off, ten thousand runners starting the 13th Berlin Marathon. It was a festive scene, electrified with life energy, the course lined with spectators, music blaring, advertisements overhead, decorations adorning the streets—all right in the shadow of the Brandenburg Gate and the Berlin Wall.

As the race began, just over the Wall, armed East German border guards, some with binoculars, others with leashed German shepherd dogs, perched on an embankment, peering in our direction. The gloomy, vacant stillness of the East lay in stark contrast to the lively, colorful party atmosphere that accompanied the start of the twenty-six-mile race. It looked, as it always did, as if two movies, one black-and-white and one color, had been spliced together, the strange contrast between a free, open society and a closed one always striking.

As I ran, Cordula and the rest of the East German women's Olympic cycling team remained secluded in Marzahn in East Berlin, a few miles from the Brandenburg Gate and the race starting point.

*U*nlike a typical marathon course, which makes a graceful loop through a city, the 1986 Berlin Marathon route was laid out entirely in West Berlin, coiling like a snake, winding around and weaving back and forth within the confines of the Berlin Wall.

Initially grouped in a dense pack, we began to spread out, making our way westward along Strasse des 17 Juni, 17th of June Street, named for the 1953 East German Workers' Uprising. We passed the Siegessäule,Victory Column, adorned by a golden statue of Victoria who stood high above at the top of a stone pillar, facing westward, as if watching over the city. I savored each step, knowing my assignment in Berlin would soon come to an end. After passing through Tiergarten and Charlottenburg, the course turned east and we ran toward Checkpoint Charlie. How many missions had our teams begun and ended at that infamous location, which had played such an important role in my life over the last few years?

As the course passed the Rathaus Schöneberg, West Berlin's main city hall, one spirited spectator held up a sign: "Ich bin ein Berliner!" a reference to President Kennedy's speech the day he stood at the courthouse in 1963 and said: "All free men, wherever they may live, are citizens of Berlin, and, therefore, as a free man, I take pride in the words 'Ich bin ein Berliner!'"

From there the route ran south through the American sector in Zehlendorf, past my apartment, and the lush green Grunewald Forest where I often jogged or splayed out a blanket, had a solitary picnic, and read a good book. The path turned again and bypassed my office at Clay Headquarters.

As I approached the last long stretch of road before the finish line, amid all the noise of the race and the crowds, I heard, "Go lieutenant!" and looked up to see some of my Flag Tour team cheering me on, one of my soldiers perched high above the street, dangling precariously atop the outstretched arm of a towering light post, which both scared and delighted me no end.

Not long thereafter, my parents came to visit me. I showed Eddie and Hanna my apartment and the compound where I worked. They knew that I worked in intelligence but nothing about what my work entailed or that I sometimes worked in the East, and they knew enough not to ask. I showed them my favorite places in downtown West Berlin.

With no way of knowing that Cordula was training in the nearby Marzahn velodrome, I took my parents to East Berlin on an official Allied Forces bus tour, a carefully orchestrated, limited tour that took authorized visitors to a few landmarks in the center of East Berlin. Once we crossed through Checkpoint Charlie, my mother immediately began to perspire, becoming terrified that, nearly forty years after she had escaped, she could still be snatched up by the Stasi and dragged back into the East. Despite my reassurances that no harm could come to her while she was in the care of a U.S. Army officer in full dress uniform, once we disembarked from the bus on Unter den Linden she became very nervous, constantly looking around for anyone who might be approaching her. After snapping a quick picture at the Brandenburg Gate, she darted for cover in the bus and

Hanna and Nina in East Berlin at the Brandenburg Gate

refused to get off again. My mother breathed a sigh of relief only once we were back on West Berlin soil.

*I*n the East there was good news for the family. Heidi was elated when she learned that Cordula was being considered for her first competition outside the Eastern Bloc, to Italy, a "capitalist country." And there was more good news. Reinhard was overjoyed when, thirteen years after he had ordered it, he finally got his new car.

*O*n one of my last Flag Tour missions before being posted back to the United States, I had a most surreal experience.

Under a pitch-black, moonless sky, in a rural part of the edge of East Berlin, our car moved through a forested area to reach a rail line. As we moved into position, hidden in the wood line but with a clear view of the tracks, I noticed a little wooden hut nearby with no lights on.

We turned our engine off. Our goal that night, a request from USMLM, was to watch the tracks for rail movement of air defense artillery systems that might be passing through East Berlin. From the woods emerged a lone, armed East German soldier whose job it was to guard that section of the tracks from curious onlookers, including us. *Especially* us. My sergeant, who had "befriended" the soldier that afternoon when he had scouted out this particular location, had given the soldier gifts of a metal U.S. Army unit badge insignia and his Bic cigarette lighter. Upon learning that the East German soldier would be there that same evening, my sergeant had promised him that he would bring more "presents."

There he stood, a young, doe-eyed East German soldier, not much older than twenty. His weapon was strapped across his chest and pointed safely downward.

My sergeant and the East German soldier exchanged greetings, even shaking hands. Then the soldier reached into his pocket and pulled out a self-rolled cigarette. He smiled a sweet smile as he handed it to my sergeant, who handed it to me as he introduced me as his boss. Then my sergeant gave the soldier a pack of Marlboros and a Snickers bar. Upon seeing the gifts, the soldier's eyes lit up and his face broke into a wide, innocent grin. He thanked my sergeant profusely, bowing up and down and patting him on the back. The young man disappeared back into the dark to his guard post and we returned to our car to wait for something to pass through on the rail line that night. We kept track of the East German soldier by the glowing tip of his cigarettes, which he smoked one after the other, no doubt imagining the repercussions of getting caught later with American cigarettes in his possession.

Sometime around midnight, we heard faint noise. It was not the train we were waiting for. It was music, and it was coming from the darkened little wooden hut nearby. Someone was taking his chances and tuning in to music from the West. And the song he was listening to was John Lennon's "Imagine."

Just before I departed Berlin for a new assignment in the U.S., the Flag Tour team invited me to a farewell dinner at a cozy, rustic German restaurant downtown.

It had always been important to me that my team understand that gender played no part in my role as team chief. But this was the mid-1980s when women were not always perceived by their male colleagues to be up to the task. And here I was—a leader of not only intelligence professionals, but also of infantrymen, the U.S. Army's tough combat troops, a male-oriented domain.

At the restaurant, we ate dinner and bantered lightheartedly about life in the office, tours on the road, and the near misses we had experienced on missions in the East. Toward the end of the evening, just when I thought things were wrapping up, a flower seller came through with a basketful of roses. I was abhorred when the senior sergeant in charge, the epitome of an infantryman, called the flower seller over and paid for some fifteen beautiful pink and red roses, which were then handed out to each of my team members.

He mumbled under his breath and I could see that they were up to something, no doubt organizing how to present the flowers to me, which would be a lovely gesture but frankly unnerved me because I did not want to finish my stint as their leader by being presented a bouquet of flowers, something they would not have done with a male boss. But I pulled myself together, preparing to be gracious about the whole thing.

They each said a few words, made toasts, some serious, some humorous, dramatically raising their roses or waggling them in my direction as they spoke, knowing full well that they were making me uncomfortable with their tributes, and especially with their flower waving.

After the last nice words were said, just when I expected them to pool their roses into a bouquet and present it to me, the sergeant in charge said, "Here's to you, ma'am," at which they all theatrically tipped their roses to me and then chomped off the heads of the flowers, exaggeratingly chewing them up and swallowing, then pooled the stems, which they presented to me.

After three years in Berlin, I chose to leave by rail instead of by air because it would be my last chance to travel through the East. I reserved a seat on the slower, less popular alternative to traveling from Berlin, the U.S. Army overnight Duty Train.

On the day of departure, a cold, wintry day, at sundown, I boarded the train at the station in the American Sector. As I settled into my compartment, the American MP crew locked and sealed the cars, a security measure required by the Soviets to prevent East Germans from trying to board so they could escape as the train made its way through the East. In Potsdam, the engine was switched out to an antiquated East German model, another requirement by the Soviets, a step which made the 115-mile trek to the West uncomfortable and painfully slow. With the crew's all-clear thumbs-up, the train rolled out of the station and slipped into the rural landscape of East Germany.

I stared out the window at the East German countryside, becoming mesmerized by the undulating movement of the train as we rolled through the barren expanse of fallow late-autumn fields, bypassing somber villages and hamlets and ashen fields of brown.

As night fell, I reflected on my time in the East. It occurred to me that I had come to Berlin in search of something. And now, it seemed, as I was leaving, in my last hours on my last day in East Germany, I was still searching.

With it pitch dark outside now, I peered out the window to see if there was anything I could discern from among the shadows. Through half-iced-over windows and the occasional assembly of leafless trees lining the tracks, I stared off into the night, seeing little more than an occasional solitary light from a house or the telltale glimmer of an outdoor bonfire.

As the train rumbled on through the frigid countryside, I couldn't help but wonder what was in store for the people of East Germany, especially as other Eastern Bloc countries were making changes. Honecker, that die-hard soldier of communism, had made it clear he would not waver in his decision to stay the course. It saddened me to think that my mother's family was destined to endure isolation and repression for years to come while citizens of surrounding Eastern Bloc nations were on the brink of enjoying a reprieve from decades of hard-line communist rule.

Somewhere around Magdeburg, I climbed into my sleeping berth and tried to make myself comfortable for the rest of the overnight ride. I read for a while, trying to clear my mind. Then I tucked in under the blanket and turned out the light, but sleep eluded me. Time passed, but the rocking of the train did not feel soothing to me and I had no desire to sleep.

At some point, the train let out a long, shrill whistle, a haunting, hollow sound that echoed far and wide, reverberating through the countryside. Years later I would learn that East Germans who had heard that piercing whistle in the night longed to be passengers on the American train headed toward the West, and toward freedom.

At around midnight, the train came to a slow, shuddering stop. We had reached the last stop in East Germany before crossing into the West, the Soviet East–West border checkpoint at Marienborn and Helmstedt.

Under the glare of bright lights, the U.S. train commander and his interpreter disembarked and stood on the platform to await a response from the Soviets. After nearly thirty minutes, a Soviet guard finally emerged from a booth. Clearly relishing his authority, he sauntered ever so slowly to meet the American train crew, took the official paperwork and identification documents of all on board, then slowly disappeared back into the booth, where he and his crew took their time to painstakingly scrutinize each document to find any mistake they could. Any minute discrepancy, no matter how small, a one-letter misspelling, a date in error, a number out of place, could cause the train to be held up for hours. It was simply another form of harassment meant to make things difficult for the Americans one last time before they left the East.

My attention could only be held by this scene for so long before my eyes were drawn to a flickering light coming from a cabin just beyond the station platform, near the forest. The window backlit, I could make out a curtain, and the silhouette of a figure, an old woman, I thought, seated, hunched over and still, like she was waiting for someone. *Someone's grandmother.*

Then a face peered out from behind the lace curtain that framed the window. Another figure in the background, *perhaps her daughter.* I could not make out the details of their faces, but my thoughts ran wild and I imagined them to look just like Oma and my aunt Heidi. I was captivated and could not look away. Though I knew it was impossible that this was Heidi and Oma, who had died nearly ten years earlier, a sense of longing rose within me and, for a brief, exultant moment I allowed myself to consider the idea. I watched the old woman in the chair by the window and envisioned Oma's face as I stared into the night, my own face reflected in the windowpane.

Down below, the Soviet guard emerged from the shed and shot a slow, furtive glance at the passenger train cars above. He returned the documents, turned his back to the U.S. crew, and walked away. We had been dismissed by the Soviets and were free to pass into West Germany.

The train began to move again. *No,* I felt like saying. *Let me stay here a moment longer.* But soon the train moved on, haltingly, achingly, on down the tracks toward the West.

23

"TEAR DOWN THIS WALL"
WINDS OF CHANGE
(1987–1988)

Who wants to talk us into changing, and why?
—*Erich Honecker*

*I*n the United States, President Reagan continued to rail against communism. In the Soviet Union, General Secretary Gorbachev worked to promote his ambitious transformation agenda. Other Eastern Bloc countries followed his lead. Honecker, however, remained unyielding, refusing to budge from the path he had set forth for his country. But with Gorbachev now gaining in global popularity, and more and more East Germans tuning in to Western newscasts, by now many had heard Gorbachev's call for change.

As part of the Intermediate-Range Nuclear Forces (INF) Treaty, Gorbachev and Reagan agreed to eliminate medium-range missiles in Europe and cut their overall nuclear arsenals in half. Not long thereafter, Gorbachev and Reagan agreed to remove Pershing and SS-20 nuclear missiles from Europe.

*I*n June 1987, President Reagan visited Berlin, stood in the shadow of the Brandenburg Gate, and made what would become one of his most famous speeches. In front of bulletproof glass protecting him from potential snipers in East Berlin, he proclaimed, "General Secretary Gorbachev, if you seek peace, if you seek prosperity for the Soviet Union and Eastern Europe, if you seek liberalization, come here to this gate. Mr. Gorbachev, open this gate. Mr. Gorbachev. Tear down this Wall."

Along with the rest of the world, I watched Reagan's address from my television at my home in Fort Meade, Maryland, where, after being promoted to captain, I was assigned to a strategic intelligence position. My mother watched the address from her home in Washington, D.C. East Germans too were tuning in, hearing Reagan's call to end repression in East Germany and end the Cold War.

In Karl Marx City, Heidi heard Reagan's translated speech from West Germany. For a fleeting moment, she wondered if Oma's prophecy about the family reuniting with my mother could ever possibly come true. But then, just as quickly, she discounted that idea, thinking there was no chance, that nothing in East Germany could change as long as Honecker, his fellow hard-line communists, and the secret police remained in control.

Before deciding whether she could join the East German national team in traveling to Italy, Cordula, now seventeen, had to undergo a special security screening.

Every year since joining the program, she had truthfully filled out a questionnaire about her connections with people in the East and West, and answered probing questions about relatives in "capitalist countries." She had always been forthright and honest about her non-relationship with her aunt Hanna, and the state minders had always seemed satisfied with her responses. But travel outside the Eastern

Bloc was an entirely different matter and they needed to be assured of the athletes' loyalties and determine that they were not interested in defecting.

The East German leadership counted on their security apparatus to root out potential flight risks. A defection would cause the regime a major embarrassment. East Germany could not afford any more negative publicity. The security official called her in.

"How often is your family in contact with your mother's sister or any relatives in the West?" the official asked, no doubt already knowing the answer.

"My mother doesn't communicate with her sister anymore."

"How would you like to live in another country?"

"I can't imagine living in another country," Cordula answered truthfully. "I have everything here. My family is here. My life is here. Why should I want to go to the West when I have everything here?"

"That's right," he said. "Your family *is* here and they are very important to you, aren't they?"

The official then told her about an athlete who had defected a year earlier.

"She had nothing when she left. No money, no support. Nothing. Now she is destitute, maybe not even alive. She disappointed her country, her family and herself. After all we did for her . . ." He shook his head, then looked at Cordula. Saying nothing further, he dismissed her, leaving her unsure about whether she had passed the screening and would be given the chance to compete internationally on Western soil.

Then, one afternoon, her trainers called her in and told her the news. She had been selected to compete in the world championships in Bergamo, Italy.

She went to call her parents, who were packing the Skoda for their weekend at the bungalow.

The instructions were crystal clear. The team had to remain as one unit, staying together at all times. No one could go off alone. Once they arrived in Italy, they could converse only with their Eastern Bloc compatriots. Talking to anyone from the West was strictly forbidden. Speaking to any West German would get one sent home and kicked off the team.

"Stay focused," the trainers reminded the girls. "You are here for one reason only and that is to represent your country proudly, and win."

With the trainers hovering over their six athletes, in July 1987 they flew to Bergamo from Schönefeld Airport in East Berlin on East Germany's state-owned airline, Interflug. On the airplane, when Cordula had to use the restroom, the trainers sent one of her teammates to go with her.

In Italy, they stayed together, moving as one tight-knit cluster wherever they went. Their trainers watched their every move.

Cordula was stunned at the differences between Italy and East Germany. She felt like she had flown onto another planet. People everywhere were smiling, even laughing openly. There was an atmosphere of openness, lightness, and unguarded happiness, the likes of which Cordula had never seen before. People looked one another in the eye, grabbed one another in big exaggerated embraces, and showered one another with enthusiastic displays of affection. There was warmth, color, and energy everywhere, in stark contrast to East Germany's subdued cold and gray.

The trainers, however, worked to dismiss the positives, constantly pointing out to the girls what they called "deficiencies of capitalist countries." When a group of nicely dressed Italian tourists stopped and pleasantly asked directions, the trainers waved them off. Turning to the girls, one of them said, "Did you see that? They were begging for money. They're homeless. They're really in bad shape." The girls snuck glances at one another, but nodded to the trainers that they understood. Cordula, however, new to the game, in her naïveté, missed the cues and blurted out, "They don't look so bad to me." Andrea elbowed her. One sharp glare from a trainer and shaking heads from her teammates and Cordula understood that she should keep her thoughts to herself.

To help motivate their athletes, the trainers gave the girls spending money to buy anything they wanted, but there were restrictions. They were prohibited from buying anything with a Western logo, and were required to keep their purchases out of sight once they got back to the East, forbidden to share or even show others what they had bought in the West. Approval to own such luxury items came with a cautionary warning not to flaunt them, so as not to risk awakening people's needs or desires.

With their allotted money, Cordula and teammate Andrea both bought Sony Walkmans, portable audiocassette players with lightweight headphones, all the rage at the time in the United States and around the world.

Thrilled with her first purchase outside the East, Cordula turned to Andrea.

"This is fantastic!" she whispered to Andrea, who smiled back at her.

After an admirable performance, the East German team returned home, the trainers relieved once they got their athletes back onto East German soil. They reminded the girls to downplay their expe-

rience and not to give anyone the impression that their trip outside
the East had been anything special, but rather that it was all business,
mundane and routine.

*O*ver the next year, Cordula and her teammates continued to compete in national and international competitions in Austria, France,
Czechoslovakia, and Bulgaria. Having medaled or placed in some of
the major races, Cordula was then selected to begin training for the
1988 Olympics in Seoul, South Korea. She and her teammates spent
the next few months in intensive training at camps and on cycling
tours. One month before the Olympics, three of the six on the national team were chosen to represent East Germany. The most junior
and the youngest on the team at the age of eighteen, Cordula was not
among the top three, but was designated an alternate.

At that Olympics, East Germany won more medals in cycling than
any other country, and came in second overall in medals, with the
Soviet Union winning 132 medals, East Germany 102, the United
States 94, and West Germany 40.

After the Olympics, Cordula remained on the national team and
continued to train and participate in international competitions.

By now, from Paradise Bungalow and at their flat in the city, Heidi
and Reinhard were regularly tuning in to news from around the
world, to track the changes occurring outside East Germany. They
listened intently to Gorbachev's speeches. The East German program
Aktuelle Kamera at 7:30 P.M. was followed by West Germany's *Tagesschau* at 8:00. While East German news downplayed the changes
taking place in the Soviet Union and other Eastern Bloc countries
and hailed progress at home, Western news media praised the sweep-

ing changes in the Soviet Union and glorified Gorbachev. Honecker remained entirely unmoved.

"Our policy of reform . . . has borne fruit and will continue to bear it," he insisted. "Who wants to talk us into changing, and why?"

Fully aware that many East Germans were tuning in to Western broadcasts, Honecker worried that news of rapidly changing events taking place in other Eastern Bloc countries could undermine his grip on power, so he tried to distract them. He adopted a strategy that gave the appearance of change by introducing lighthearted comedies and lively films with toned-down propaganda, in a more subtle campaign to discredit the need to reform.

*I*n 1988, at the height of his fame and in the middle of his Bad World Tour, Michael Jackson gave an open-air concert in West Berlin, on the Reichstag lawn with the Berlin Wall as a backdrop. East German authorities, knowing full well that their young citizens would be drawn to his music, staged a competing rock concert in East Berlin, emceed by East German Olympic figure skating superstar Katarina Witt. But that didn't stop some from trying to get close enough to the West to hear the King of Pop.

At first, East German security forces did nothing, even appeared to allow it, but when more people started streaming in to get close to the Wall, suddenly everything changed. The VoPo brought out their truncheons and violently descended upon the crowd. As concertgoers on the west side of the Wall were having the time of their lives, dancing and singing along to "Thriller" and "Man in the Mirror," under the cover of darkness on the other side of the Wall, hundreds of young East Germans were being clobbered and hauled away.

In the mid-1980s, after Roland retired, he started writing more frequently. Though his words were carefully chosen, his letters vague and lacking in any real information, Hanna was deeply happy that she could finally be in touch with her much-loved big brother, whom she had been completely robbed of a connection with and had missed so terribly.

But in 1988, shocking news came from the East. Roland's wife wrote to say that Roland had died. Hanna was devastated. She had not even known he was ill with diabetes. He was only sixty-two.

Roland's wife and son and the entire family gathered to bury him in Valfitz, where he had served proudly as a school director and teacher for so many years.

With all the changes taking place in Eastern Europe, Heidi and many others couldn't help wondering what would happen in East Germany if Honecker was the only Soviet bloc leader to resist change. She thought about the Workers' Uprising of 1953 and could not imagine a revolution in East Germany without bloodshed.

Heidi looked at photos of our family, my parents and us six children. In the United States, my mother looked over photos of her family in the East. She watched Gorbachev on television and tried to imagine what it would all end up meaning for East Germany and ultimately for her family. We were soon to find out that, though Gorbachev had little influence on the East German leadership and though Honecker remained entrenched, there would indeed be change.

In fact, the winds of change were already blowing without Honecker's consent.

24

"GORBY, SAVE US!"
A NATION CRUMBLES
(1989)

If not me, who? And if not now, when?
—*General Secretary of the Soviet Union Mikhail Gorbachev*

*I*n the East, Warsaw Pact nations responded to Gorbachev's call for change. In Hungary, the parliament voted to allow freedom of association and assembly, to permit the establishment of political parties, and to set a date for multiparty elections the following year. This daring move when tried thirty years earlier had provoked a brutal Soviet crackdown. This time Moscow did not intervene.

In Poland, the Solidarity movement, striving to throw off the shackles of communism and create a democracy, continued to gain momentum. Honecker still refused to budge.

*I*n February 1989, a twenty-year-old East German named Chris Gueffroy, whose greatest wish was to see America, believing that border guards had been ordered to stand down their shoot-to-kill orders, tried to escape over the Wall. No such order had been issued, however, and border guards shot and killed Gueffroy.

One month later, Winfried Freudenberg made it out in a home-made hot-air balloon but was killed when it crashed to the ground in the American sector in Zehlendorf, not far from the apartment I had lived in for three years.

By June, in a crushing defeat of Polish communism, Lech Walesa's Solidarity Party swept the elections. When the Polish communist authorities phoned Gorbachev to find out what they should do, if they should really concede and abide by the election results, Gorbachev said, "The time has come to yield power."

In China, hard-line leaders called in the army to suppress a democracy movement in a brutal and bloody crackdown in Beijing's Tiananmen Square. Honecker watched it all, envisioning a Tiananmen-like crackdown should things get out of hand in East Germany. However, he wondered how the lack of Soviet backing would affect his ability to retain control.

For most East Germans tuning in to the transformations taking place all around them, it was life as usual, especially in light of the fact that only a few months earlier, Honecker had declared, "The Wall will still be standing in a hundred years."

For Heidi and Reinhard, life went on. At work, no one spoke of revolution or the possibility of change in East Germany. The rest of the family too went on with their lives, teaching and working in the towns and cities in which they lived.

Cordula and the national team continued to train, compete, and win races at home and abroad.

In June 1989, Cordula became the East German national women's champion in the points race event, a mass cycling event involving dozens of top East German racers covering some one hundred laps.

In July, Cordula and the East German national women's cycling team flew to Switzerland.

By now a seasoned team member, she knew the routine.

The girls dressed into their blue-and-gray competition uniforms in the locker room, then emerged onto the track as one unit. As their trainers lined their bikes on the track, the girls readied themselves, stretching and listening to their trainers' last-minute words of motivation.

Suddenly there was a terrific scuffle. All the East German trainers bolted from the track and disappeared, leaving the team alone to finish setting up without them.

With the start of the race only seconds away, team members positioned themselves and tried to focus. But then one of the girls asked, "Where's Andrea?"

Looking around, they noticed Andrea was not among them. The race began, the other cyclists took off, leaving the East German team struggling to refocus.

At the end of the race, one of the trainers stood waiting for the girls. The other trainers did not show up until hours later, and when they did, there was no explanation.

The next morning, they told the girls the news. Andrea had defected.

Cordula and the others were astounded. It was incomprehensible and completely unexpected. Andrea had been a model East German athlete and had never let on about her intentions.

And so the East German national women's cycling contingent returned home without one of their strongest, most capable, and most talented members.

Less than one month later, the dominoes began to fall.

By August 1989, Hungary had disabled its border defenses, including electrified fences with Austria, effectively opening passage into Western Europe. Word of the opening to the West spread via media reports. By September, more than thirteen thousand East German "tourists" in Hungary had fled to Austria. Honecker called them scum and ingrates who had abandoned the cause.

Inside East Germany, many were driven to act. From a small gathering at a Prayers for Peace meeting at the little Nikolai Church in Leipzig, a demonstration of about a thousand kicked off but was quickly suppressed by the VoPo. But instead of dispersing, more demonstrators streamed in, reviving the protest, and before long, tens of thousands were taking to the streets in that city, overwhelming police and calling for peaceful, democratic order in East Germany and chanting in unison, "We want out!"

By mid-September 1989, in the United States, I was glued to the TV in my home in Maryland, carefully following the extraordinary events unfolding in Eastern Europe. Now married to a fellow army officer, who was also a U.S. Army Russian Foreign Area Officer, I was nine months pregnant with my second child as my husband and I watched thousands of East Germans making their way to Poland, Hungary, and Czechoslovakia and those seeking asylum on the grounds of West German embassies.

In Washington, my mother too watched as families who had driven from East Germany abandoned their Trabants in the streets of Prague and scrambled over walls to get onto the grounds of the West German embassy, passing babies, young children, suitcases, and strollers over the back fence of the embassy compound and into the welcoming hands of fellow East German refugees who had already made it over and were waiting on the other side.

Over the next days and weeks, that crowd swelled into the thousands, camping out on the lawn, worrying and waiting for some sign that the West Germans would not send them back to East Germany, where they would face dire consequences as traitors to the regime.

On September 30, their fate still uncertain, West German foreign minister Hans-Dietrich Genscher appeared on the embassy balcony and looked over the crowd. Terrified but hopeful, on pins and needles, East Germans stood waiting in complete silence, hanging on his every syllable.

"Dear fellow Germans," he began slowly. "We have come to you in order to inform you that today, you are free to leave for West Germany. . . ." But before he could finish, the crowd erupted into a thunderous uproar. Wild, elated cheers drowned out Genscher's address as people burst out crying and hugging one another. Genscher concluded by directing families with babies to be on the first train bound for freedom in West Germany.

With Poland and Hungary grabbing their freedom, and East Germans beginning to rebel, there was no indication from Honecker that he would relinquish his reign and release his people. In the United States, my family assumed that the likely outcome would be a violent crackdown.

In Karl Marx City, Heidi and Reinhard watched West German newscasts of the refugees spilling out through Hungary. Like us in the United States, they too followed the whirlwind developments taking place in Austria, Poland, and Czechoslovakia, not fully comprehending or even believing that any of it was actually real.

I protected myself from being overly optimistic, reluctant to think more than forty years of totalitarianism could simply disappear overnight.

*W*ithout an invitation from Honecker, Gorbachev visited East Germany and took to a public stage. A very pensive Honecker stood by his side. The people of Eastern Europe, he said, had a right to choose their own futures. Honecker wore a blank stare but Gorbachev's words set the crowd ablaze.

They began to chant, cautiously at first, their calls quickly gaining momentum. "Gorby, Gorby, Gorby!" they yelled louder and stronger, the cadence picking up as more in the audience chimed in and the chant grew more defiant.

"Gorby, Gorby!" With the support of the leader of the Soviet Union seemingly now on their side, the crowds were suddenly unafraid of their own leader and they continued to drum in unison, as some voices cut above the din, yelling, "Gorby! Save us!"

*W*ith much of East Germany wondering how or even if their leaders would mark the fortieth anniversary of the founding of East Germany, Honecker refused to be sidetracked. The stage was set as Honecker prepared to mark forty years of communist progress in East Germany, hail the country's achievements, and promote his plans for the future.

On October 7, 1989, as people were fleeing through Hungary and Czechoslovakia, Gorbachev arrived in East Berlin, having agreed many months earlier to be the guest of honor for East Germany's biggest and most important celebration ever.

In Karl Marx City, the festivities kicked off with their homegrown champion, Cordula, and her home team participating in a race down the center of the city. No sooner had celebrations begun when spontaneous demonstrations erupted. VoPos descended. Convinced that violence would ensue, Cordula escaped the explosive city center and rode to her parents' apartment to wait for Heidi, Reinhard, and Mari to return home.

Over the next days, Cordula, her parents, and Mari watched news events on Western television as police all over East Germany tried to push back demonstrators with physical force, using batons and water canons. As East German marchers recalled the 1953 uprising, many realized that this time things were different. They felt empowered knowing they had Gorbachev's support and the West on their side, and that they were riding the momentum of other Eastern Bloc countries in transition. Still, watching the events unfold on television, Cordula, Heidi, and Reinhard could not help but wonder where it was all headed and how it would all end.

People throughout East Germany took to the streets, disrupting anniversary festivities everywhere. With disorder now spreading, Honecker, acting as if nothing out of the ordinary were happening, declared the show must go on.

On a crisp, chilly day in East Berlin, Karl Marx Allee was lined with thousands of spectators holding flags as they stood in front of six-story-high billboards proclaiming *"40 Jahre DDR"* (Forty Years of East Germany).

In the VIP box, Soviet leader Gorbachev took his place next to Honecker, who wore a strangely beatific smile as he cast his hand in a royal wave over the assembled masses. The main streets, topped out in full-propaganda blossom, were lined with enormous red banners that whipped in the wind. With speakers in place, dignitaries seated and attentive audiences prepared to celebrate, the parade began.

A bell sounded, signaling the start of ceremonies, and spit-polished military and paramilitary personnel marched. Behind them, bedecked FDJ and Young Pioneers, bands, and other communist groups followed as the crowd looked on and cheered. Military equipment rolled past.

Over the next few hours, as hundreds of demonstrators attempting a peaceful march on the outskirts of the gathering were grabbed by the VoPo and shuffled away, the two leaders, oblivious to the mayhem, stood side by side, watching the parade. In an eerie state of denial, Honecker rattled on, regaling an annoyed Gorbachev with delusions of his country's achievements, affirmations that now only few believed. Not long after the parade, Gorbachev would tell Honecker, simply, "Life punishes those who delay."

Despite the unrest taking hold, and euphoria sweeping the country, the great majority of East German citizens were still completely perplexed and chose to keep their distance from the turmoil. Many chose not to take part in demonstrations, believing it could only mean trouble for those who did.

Many, including Heidi, still believed the events were staged, part of an elaborate hoax intended to root out the disloyal. Even if the demonstrators gained momentum, they thought, the state was all-powerful, and any act of rebellion was doomed to failure. To many, the whole thing felt like a twisted psycho-thriller film in which the entire country's population was the cast, so they stayed away, trying to follow news of the events from the safety of their homes.

From Leipzig to Dresden, demonstrations broke out as hundreds of thousands of people now marched, calling for change.

In Leipzig alone, 300,000 citizens with banners affirming "We are the people!" marched in the streets. They stormed past the local Stasi headquarters, shouting their hatred for the regime, demanding an end to secret police control. Honecker called for the protest to be put down immediately, and the Stasi chief ordered his forces to suppress the mobs. But as police and local authorities watched the mass of demonstrators advancing on the Stasi headquarters, they were overwhelmed and chose not to intervene. In East Berlin's Alexanderplatz, the crowd swelled to half a million, carrying signs that read, NO LIES, NEW PEOPLE, and FORTY YEARS IS ENOUGH!

25

THE WORLD IS STUNNED
"SCHABOWSKI SAID WE CAN!";
OR, THE WALL FALLS
(November 9, 1989)

What is right will always triumph.
—*President Ronald Reagan*

*J*ust eleven days after Gorbachev uncomfortably joined Honecker for the forty-year celebration of East German rule, an event that was supposed to mark a shining achievement for the Honecker regime, the East German Politburo, in an effort to reinstate calm, and with Gorbachev's approval, forced Erich Honecker from office.

Honecker was replaced by his deputy, Egon Krenz, who tried to quell the crowds and get things back under control. Krenz called on the population to remain calm, promising change in East Germany. But it was too late.

Across the country, people took to the streets. In East Berlin alone, the numbers of demonstrators marching peacefully for change swelled to an estimated one million, remarkable for a country of 16 million.

*I*n early November, Cordula and the national women's cycling team were back training at Marzahn in East Berlin.

"You might notice things happening in the streets of Berlin," the trainers told the girls, "but that is none of your concern." With that, the athletes promptly returned to training, made to focus solely on timed trial cycles, speed work, and strength exercises.

*I*n Karl Marx City, Heidi and Reinhard watched in awe: the demonstrations, the crackdowns, the chaos, people fleeing the country. Opa's prophecy about the regime falling apart seemed to be coming true.

In early November, the East German leadership finally took to the airwaves to address the citizens of the East about the chaos. Heidi and Reinhard and the great majority of East German citizens were utterly unprepared for the earth-shattering announcement their government would make that day. It would change their lives forever and alter the course of history.

*A*larmed at the hemorrhaging of East Germans through the now-open borders in Hungary, the Krenz administration scrambled to come up with a plan. With East Germany's very existence now at stake, Krenz called the country to order and promised a loosening of travel restrictions, saying that, with the proper permissions and paperwork, citizens would find it easier to travel outside of the East.

Günter Schabowski, a Communist Party official, was given the task of relaying Krenz's policy to the press and to the public. At a quickly assembled international press conference on November 9, before a roomful of reporters, Schabowski, who had not been present at the meeting with Krenz, mixed up the original intent of the new policy when he stated, "The regime has decided to invoke a ruling to allow every citizen of the East German republic to emigrate through East German border posts." In essence, he essentially mistakenly conveyed that all travel restrictions were being lifted, making it sound like the citizens of East Germany were free to leave.

But then Schabowski halted, seemingly perplexed about what he had just read, unsure as to what exactly the order really meant or if he had gotten it right.

Heidi and Reinhard looked at each other.

When a reporter, believed to be NBC journalist Tom Brokaw, asked when the decree would go into effect, a confused-looking Schabowski answered, "As far as I know, immediately."

In that one instant, the order that promised to relax travel procedures dissolved them.

Within a half hour of the broadcast, West German news announced that East Germany had opened its borders.

Puzzled and suddenly feeling light-headed, Reinhard turned the television channel to West Germany's ARD news just as the anchorman began.

"This is a historic day," he said. "East Germany has announced that, starting immediately, its borders are open to everyone. [T]he GDR is opening its borders. The gates in the Berlin Wall stand open."

It didn't take long for East Germans to rush the checkpoints between East and West, including Checkpoint Charlie, demanding that border guards immediately open the gates. The guards, unsure of what to do, made frantic telephone calls to their superiors, who ordered them to finger the "most aggressive" people at the gates and mark their passports with a stamp revoking their citizenship and barring them from ever returning to East Germany. The guards turned back to face the crowds of tens of thousands who now demanded to be let through, calling "Schabowski said we can!" and rattling the fence, shouting, "Open the gates! Open the gates!"

Border personnel, greatly outnumbered and completely overwhelmed, had no way to hold back the huge crowds and finally gave in, opening the checkpoints and standing back, allowing the crowds to flee through to the West. As throngs of East Germans poured through the gates, they were greeted by West Germans waiting on the other side of the Wall with flowers, champagne, and even money, giving pats and hearty brotherly embraces and greetings of "Welcome to West Germany!"

Heidi watched the scene unfold on television, saw people dancing on the Wall, but simply did not believe any of it was real. Reinhard was confused. Even as they watched television coverage of the masses of people streaming through the borders, Heidi trusted none of it. Having lived her entire life within the East German system, her defenses remained sharp and she was convinced it was a fake newscast, part of an elaborate ruse orchestrated by the East German secret police to gauge citizens' allegiance and ferret out enemies of the state.

Even when she heard loud music, elated voices drunk with happiness, and cheers streaming out from her apartment complex, from up above and down on the street below, Heidi remained sure it was just a matter of time before the regime would crack down and all would become as it had been before. In Salzwedel, Dresden, and Naumburg, Manni, Tiele, Helga, and Tutti watched in awe.

*I*n East Berlin, still sequestered in the sports hall, Cordula and her teammates were oblivious to the masses leaving the city just a few miles away; their trainers demanded their complete focus on training. That evening they directed the girls to turn in to bed early so that they would be fresh for the next day's workout. Once the girls were asleep, the trainers slipped out of the building and took off to see for themselves what was happening at the Wall. They crossed into the West and partied with the thousands of others who had gathered; by early morning, as the sun was rising, they returned back to the sports hall.

That morning, in the pale light of a new day, Cordula awoke to a knock at the door, a trainer telling her and the other girls to get up and assemble immediately.

Once the team had gathered in the common room, the trainers, with wide grins, stood before them, barely able to contain their joy. One of them said, "The border is open."

They looked at each of the girls in turn, waiting for their reaction. But the girls did not react. They looked at one another. Cordula wondered what kind of a joke the trainers were playing. Were they testing the girls for their loyalty, as they sometimes did? The trainers took the girls to a television and turned to Western coverage of the masses at the border and people atop the Wall.

The girls stood silent, shocked and in complete disbelief.

"We were out there last night," one trainer said excitedly pointing to the scene playing out in West Berlin.

"Wait," Cordula said, suspicious, knowing even a quick unauthorized trip to the West could get one tossed in prison for years.

"What does that even mean, the borders are open? For how long?"

"For *forever!*" yelled the trainer into Cordula's face, his intense stare boring right through her. "Listen to what we are saying to you!" he pleaded. "*The border is open! We are free!*"

The girls reacted differently to the news. Some of them were elated, wanting to waste no time in joining the throngs investigating their newfound freedom in West Berlin. Others, like Cordula, were more cautious. She tried to process the information rationally. What did it actually mean? How would it impact her and her family, her life as an athlete in the short term and in the long term, and what would if mean for the future of all East Germans?

It was, by any measure, an extraordinary day. In an instant, the Berlin Wall, the impenetrable concrete barrier that had divided Germany for decades, simply ceased to exist.

In the blink of an eye, 16 million East Germans were finally and unexpectedly set free.

Within a matter of just a few hours, the people of East Germany suddenly found themselves citizens of a country and an ideology that no longer existed. Tens of thousands rushed the border to the West in a combination of disbelief, euphoria, and anxiety, not quite sure what they would find on the other side. Where just hours earlier, anyone daring to attempt to scale the Berlin Wall would have been shot or hauled off to prison, all of a sudden thousands stood atop it, celebrating as others took sledgehammers and chisels to it, trying to destroy the world's most recognized and hated symbol of Cold War oppression.

In the end, amid all the changes that were rapidly being unleashed by the actions of Soviet leader Mikhail Gorbachev, the all-powerful East German state, a regime that had tried to control and claim responsibility for every aspect of its citizens' lives, didn't have the answer for its own survival.

In the United States, no one in my family could fathom that it was true. By now the mother of a toddler and a newborn, I was astounded as I watched the news from the sofa in our apartment in Fort Meade, Maryland: people dancing, drinking champagne, partying on the Berlin Wall, on the Ku'Damm, and throughout West Berlin.

In Washington, my parents, Hanna and Eddie, watched in silence, staring dumbstruck at the television.

"I'm Peter Jennings in New York. Just a short while ago, some astonishing news from East Germany where the East German authorities have said in essence that the Berlin Wall doesn't mean anything anymore. The Wall that the East Germans put up in 1961 to keep its people in will now be breached by anyone who wants to leave."

Forty years after my mother had escaped, the thought of finally being able to see her family again overpowered her. Dazed, she sat down to try to process it.

Heidi was the first to call. From Karl Marx City, she phoned my mother in Washington. The sisters, at first overcome with emotion, greeted each other gently, haltingly. It was a bad phone connection, but even through the buzzing, fading signal, they managed.

Heidi told Hanna that she was confused, that she could not fathom that East Germany had really fallen and no longer existed. Like a big sister comforting her little sister, my mother reassured her that it was indeed true. When Heidi said she had still not been to West Germany to have a look around, my mother encouraged her to go, to take just a short trip at first, so as not to become overwhelmed.

26

DAWN
LEAVING THE EAST
(Autumn 1989)

Remember tonight . . . for it is the beginning of always.
—*Dante*

*I*t took yet another couple of weeks for Heidi to fully absorb that East Germany had indeed fallen, and to comprehend that she was really free to go.

One weekend in late autumn, on a clear, crisp morning, Heidi and Reinhard got up at sunrise and packed the Skoda for a daylong trip. Instead of going to the bungalow that weekend, they set out from Karl Marx City and drove toward the East–West border.

Born into the East German communist system, they had spent their lives behind the Iron Curtain, told that the West was filled with criminals, warmongers, and evildoers bent on destroying them.

Excited and nervous, they approached a crossing near Grosszöbern. They fell into line behind other motorists slowly making their way westward in their Trabants and Wartburgs. Some had pulled off to the side, anxiously waiting, worrying, wondering about what lay up ahead, afraid to leave the only life they had ever known.

Heidi and Reinhard drove slowly past, looking at them as they stared back, some with tear-filled eyes, bewildered and unsure. As the two moved onward, trepidation melted into nervous excitement, and then, when they saw that there were no armed guards to keep them from leaving, their fears gave way to jubilation. They moved through the abandoned border post, obstructed by nothing and impeded by no one.

Once on the other side, they drove for a mile or so. Then they pulled off to the side of the road, stopped the car, and just sat in silence, gazing at the horizon. Adjacent to the road was a farmer's field laid bare by the autumn harvest. They got out of the car and walked alongside it, breathing in the fresh air. Off in the distance, a farmer, no doubt aware of where they were from, waved a welcome.

After a time, they walked back to their little car and got back in. Reinhard started up the engine. She looked over at him. He smiled at her, and they continued on their way, driving onward into the beautiful unknown.

27

REUNION AND REBIRTH
TOGETHER AGAIN
(1990–2013)

I may not live to see the day, but you will be
reunited with Hanna.
—*Oma in Klein Apenburg*

*I*n the East, the extended East German family held a meeting
about how best to proceed with their lives, and to discuss reconnecting with Hanna.

In the spring of 1990, more than forty years after she had last seen
her family, Hanna, now sixty-three years old, and Eddie flew to Germany to reunite with her family.

Only Manni and his wife came to meet Hanna at the Frankfurt
airport, in part not to overwhelm Hanna, but also because they could
only fit four people and a suitcase into Manni's Trabant.

Manni smiled broadly through his tears when he saw his sister,
and greeted Hanna with a bouquet of flowers.

She embraced him, seeing how the years had lined his face.

Strands of silver streaked his dark hair. He gazed back at her. She was shorter than he remembered her, and her once-striking features had mellowed to soft creases, but she still had the same smile, their mother's smile. She guarded her emotions until he let loose his own feelings. Nothing was said, but they held each other for a long time. Manni, only thirteen years old when she had last seen him, was now in his mid-fifties.

They arrived at Manni's to a house full of people. A moment many years awaited, this was the day they had always hoped for but had never fully dared dream would happen.

Everyone stood outside waiting for her.

Hanna emerged from the car and was met by Tiele, Helga, and Tutti, who stared back at her. All of them felt the moment was hardly even real, but then they fell into one another, hugging and unleashing years of suppressed emotion.

There were so many feelings all at once: great excitement and wonderment, immeasurable joy, but there was also heartache for those who had passed, who had not lived to see this day. There was anger at the system that had severed their family. And there was sadness for the time that had been stolen from them, a deep, primal pain for all that had been lost in the forty years since they had last seen one another. But there was also great comfort in the belief that Oma was with them, especially at that moment, as they finally came together once again.

*H*er sisters released her and stepped away so that Hanna could move on to greet the others.

She barely recognized her youngest sister. Hanna gazed at Heidi, who was already in tears. She was a grown woman, tall and statuesque, but all Hanna could see was the face of that little girl with long braids who had loved Heidelberg, the little girl she had only met once for a short visit nearly four decades earlier.

There was not a dry eye in the room when Heidi and Hanna came together—and especially when Heidi took Hanna's hand and didn't let go for the rest of the afternoon.

*H*anna and Eddie met Reinhard and Cordula and many other relatives that day and over the next few days. She renewed her relationships with her brother and sisters, met their spouses, children, and grandchildren. She also met Roland's, Klemens's, and Kai's families.

That evening, everyone came together to formally celebrate an extraordinary reunion.

They made toasts, praising their parents for their commitment to creating a closely bonded family despite their challenging conditions. They paid tribute to Opa's moral courage, tenacity, and unrelenting will in speaking out against injustice, and in his struggle to live the truth. And they celebrated Oma's spirit, unbending faith, and determination to protect the family against destructive influences; her force had sustained everyone through years of isolation and separation. And finally, the family honored my grandmother as an angel who still, always watches over them.

There was immense joy in those first days, celebrated with plenty of Rotkäppchen champagne at every turn. At times they were too overwhelmed to celebrate or even talk, though they constantly hugged one another and cried a lot. A muted melancholy hung over those first meetings, all of them wanting to strive to create a new sense of normal and in so doing, try to guard against raw emotion that could only bring sadness and hurt.

A few days later, Hanna's five siblings accompanied her to the cemetery to see Oma's and Opa's graves. They gave her room to approach on her own and in her own time. She laid flowers at the base of their gravestones and took a quiet, private moment to "speak" to each of them.

They visited Roland's, Klemens's, and Kai's graves as well, my mother particularly saddened that Roland had died only one year before the Wall came down.

In a caravan of Trabants, Wartburgs, and a Skoda, they went on to Klein Apenburg to look over the little hamlet and the house that had been Oma and Opa's home while in internal exile. Walking the grounds, Hanna surveyed the patch of earth that had once been Oma's lush garden and her oasis of peace. Then she went to sit for a while on the "Opa's resting place" bench.

The next day, they traveled to Seebenau to see Kallehn's house, his farm, and to look over the area where she had made her initial failed escapes not long after the Soviets took over the East. In Seebenau, she recalled Kallehn's unwavering support in urging and then helping her to flee. They took her to the area where she had made her final escape.

Hanna and Eddie spent the last days of that first reunion with Heidi and Reinhard at their apartment in Karl Marx City. On their last day together, they took a trip to the countryside, where Heidi and Reinhard wanted to show Hanna and Eddie a place that had become special to them during the Cold War years, and continued to be even after the Wall fell. They parked the car on a rural side road and the four got out.

They made their way down a dry dirt path, passing trees and fences that cordoned off divided plots of land. Halfway down the road, Reinhard stopped and unlocked a bolt on a wrought-iron fence post, opened the gate, and Hanna stepped into Paradise Bungalow. That afternoon, they sat in the shade of the blue spruce and pear trees, and began to get to know one another.

*I*n 1991, I went back to Germany with my husband and children and met everyone for the first time. It was extremely exciting but also overwhelming to go from having no relatives to suddenly having so many.

I met them all, including Cordula. It would be years after the fall of the Wall that we would learn so much more about one another and about our parallel lives on two sides of the Iron Curtain. Over the years, we have formed a bond that has developed into a special friendship.

Then in September 2013, I flew to Berlin with Hanna and my brother, Albert, for the latest family get-together, and to run the Berlin Marathon again. Almost thirty years after I had first run the marathon, I ran the race again, this time made epic when I ran with Cordula and Albert on a race course that holds no trace of the Wall and runs through a reunited, open, and free Berlin, finishing at the world-renowned symbol of unity, the Brandenburg Gate.

The 2013 reunion party was held in the Eberswalde Forest, thirty miles northeast of Berlin in a log cabin that had once been used to house East German youth on overnight FDJ scouting trips. More than sixty family members from the United States and former East Germany attended.

Inside, the cabin was awash in a golden glow of a rustic chandelier. The music was loud and partygoers were scattered densely throughout the room, greeting one another with big embraces and heartfelt hugs. Beer bottles tapped and champagne flutes clinked one another.

On the dance floor, Tiele danced with Reinhard, Manni with his wife. Kai's children greeted my brother Marcel, and Helga and Tutti chatted with my sister Maggy. Manni's boys joked about something with Klemens's daughters, my children, and Cordula, which caused everyone to break out in laughter. Mari's and my sister Sachi's children played together; nearby, Albert and Roland's son were deep in conversation.

At the head table, my mother and her siblings seated themselves. When the sound system began to blare an old German folk song they instantly recognized from their childhood, they quickly locked arms, linking themselves in a chain, beaming at one another and giggling like little children as they began to sway animatedly from side to side in rhythm to the music. Melting into their own private world, the six seemed oblivious to the rest of us as they sang loudly and blissfully with wild abandon. Though East Germans had been set free nearly twenty-five years earlier, these siblings still celebrated as if the Berlin Wall had fallen that very day.

Watching the family together, my mind began to wander and I stepped away from the celebration to take a break, to immerse myself just for a moment in my own uninterrupted thoughts. I panned the room, taking it all in, when my thoughts drifted back to the Cold War. In 1985, the mere thought of this kind of celebration would have been a reckless fantasy.

On the fireplace mantel stood two silver-framed black-and-white portraits. One was a picture of Opa in his middle years, looking proud and dignified. He would have been pleased with this day, I thought, to know that his children and grandchildren were all right.

The other photograph was a picture of Oma. It was the same one I had seen when I first set eyes on her as a little girl. She looked serene

Six of nine remaining siblings reunite. (*Left to right:*) Helga, Tutti, Manni, Hanna, Heidi, and Tiele.

and wise and every bit the perfect grandmother as the first time I saw her. But where once, in a little girl's fantasy I imagined she was looking at me, now she seemed to me to look out over the whole room as if to say, *My children, my family, now you are all together . . . just as I knew you would someday be.*

EPILOGUE

*O*ma, Opa, Roland, Kai, and Klemens died in East Germany during the Cold War. Manni, Tiele, Helga, and Tutti all retired as teachers. After the fall of the Berlin Wall, many East German teachers, including many of my relatives, were laid off when Germany suddenly had a surplus of educators.

Today, Hanna's brothers and sisters, their children, and their grandchildren have adjusted to life in a reunited Germany. As Oma had always insisted, today they remain very close with one another and with Hanna.

Hanna and Eddie raised six children. Now widowed, Hanna still works as a German teacher, writer, and painter. She recently published her first novel, *Christine: A Life in Germany After WWII*. She has fourteen grandchildren. Since her escape in 1948, she has been thankful every day for her freedom and, since 1989, for reunion with her family. Eddie passed away in 2008.

Albert, a retired U.S. Army colonel, now works as a senior China analyst.

After the Wall fell, Heidi's firm was privatized by Rawema. Heidi, who had never joined the Communist Party, was the only employee besides Meier, her boss, who was asked to stay on. She was promoted to executive assistant to the company president. She recently retired after thirty-seven years with the company.

Siemens privatized the company that Reinhard had worked for, incorporating it into their own electronic control systems in Erlangen, West Germany, and inserted their own West German employees. Consequently all thirty former East German employees, including Reinhard, lost their jobs. Though opportunities dried up quickly for many East Germans after the fall of the Wall, Reinhard, having so ingeniously built the bungalow during the Cold War with any resources he could find, ironically was hired as a salesman for building materials.

Heidi and Reinhard still enjoy Paradise Bungalow and the fruits of freedom every day, including as much travel as possible outside the country.

Hanna and Heidi remain very close. In 2015, Hanna turned eighty-eight. Whenever they come together, Heidi, now sixty-seven, still scrambles to hold her big sister's hand as if she never wants to let go.

Cordula continued training and competing in international competitions even after German reunification. She was the last East German champion in point track races. In April 1990, she competed for the first time in a united Germany. Competing alongside was Andrea, the defector, who had by then become a member of the National Cycling Team of West Germany.

Summer 1991 saw the first National Championship of Germany, with East and West cyclists competing together. In a surprise finale, Cordula and her team from Sportsclub Chemnitz (formerly Karl Marx City) won the race, becoming the first champions in a major cycling tournament in newly reunited Germany. In April 1992, Cordula competed in her last race as a professional athlete. Today Cordula is a deputy bank manager.

HISTORICAL NOTES
After the Fall

The unofficial dismantling of the Wall began almost immediately and, in the days and weeks that followed the opening of the borders, *Mauerspechte*, or "Wall woodpeckers," people armed with sledgehammers and picks, appeared on the scene to chisel and chip off bits and pieces of the Wall, some destroying it, others hoping to preserve bits of history as souvenirs. Over the next weeks and months, bulldozers demolished parts of the Wall, then rebuilt and reopened roads and transportation routes between East and West.

By December 1989, the Wall at the Brandenburg Gate had been dismantled and was once again opened for through traffic. In March 1990, East Germans voted in free parliamentary elections for the first time, with the Christian Democratic Union, a noncommunist party, winning the most seats in Parliament. In May 1990, the two German states signed a treaty agreeing on monetary, economic, and social union and by October 1990, East and West Germany had reunified.

*P*rofound change took place throughout Eastern Europe. After 1990, all of Eastern Europe's former communist regimes were replaced with democratically elected governments. In Poland Solidarity leader Lech Walesa and in Czechoslovakia playwright Vaclav Havel, both dissidents, were elected president of their respective countries.

In Romania, Nicolae Ceausescu, the only hard-line communist besides Honecker to reject reforms, ordered a violent suppression of demonstrations, then fled the country. He was caught and returned to Bucharest, where, along with his wife, he was executed.

Honecker fled to Moscow.

In late 1991, the Soviet Union dissolved, which officially marked the end of the Cold War.

What Happened After the Fall?
The Wall
Today parts of the Berlin Wall are on display in 140 countries throughout the world to remind people of the danger of totalitarian regimes. Concrete sections are located throughout Eastern Europe, in Gdansk, Poland; Budapest, Hungary; Moscow, Russia; in the United States at, among other locations, the Ronald Reagan Presidential Library in Simi Valley, California, and the Pentagon; and throughout the Western world, in the United Kingdom, Germany, and in South Korea, not far from the North Korean border.

The East German People
East Germans emerged to face and adjust to a new life. While some embraced their freedom, others remained melancholy, feeling a sense of loss, and some were fearful of the uncertainty that lay ahead. Most would agree, however, that they were bewildered at the extent to which their government had betrayed them.

Germany's Leadership Today
Today, Angela Merkel is chancellor of Germany. A former East German research scientist and the daughter of an East German pastor, she was one of thousands to cross into West Berlin in the initial hours after the fall.

Merkel, the first female chancellor, has led a reunited Germany since 2005.

Joachim Gauck, formerly an East German Protestant pastor, was an anticommunist civil rights activist during the regime. Today Gauck is president of Germany.

East German Leader Erich Honecker

Honecker and his wife, Margot, fled to Moscow to avoid prosecution on charges of Cold War crimes. In November 1990, he was tried in absentia for manslaughter for ordering border guards to shoot East Germans trying to escape. Following the dissolution of the Soviet Union in December 1991, Honecker took refuge in the Chilean embassy in Moscow, but was extradited by the Yeltsin administration to Germany, where he stood trial but was released due to ill health. He moved to Chile and died of cancer in 1994 while in exile, unrepentant and successfully evading prosecution for human rights abuses committed during his regime.

The Stasi

At the height of their power, the Stasi had employed one informant for every sixty-six residents; factoring in part-time informants, the number more accurately approximates one in six East German citizens.

During the chaos of the downfall, enraged citizens stormed and overran Stasi headquarters in East Berlin and began destroying files, but that emotionally charged activity was halted when they realized that the documents would be needed to prove secret police activities.

In 1992, Stasi files became accessible to the public. People could now read their own dossiers and find out who spied on them. By 1995, many documents, including those that had been shredded or otherwise destroyed by the Stasi or angry citizens who had converged on the headquarters, had been painstakingly reassembled, revealing millions of crimes and systematic human rights abuses.

Most Stasi officers were never charged. Today they live among the general population, many having gone on to find other jobs and becoming active members of society in a reunited Germany. After the fall, some 85,000 full-time Stasi officers lost their jobs virtually overnight.

The VoPo, the People's Police

Following reunification, the East German police were required to fill out questionnaires concerning their political and professional history before being accepted into the reunified police force. Those accepted were retrained and paid a fraction of what their Western counterparts received. As was the case for many employers considering former East Germans for jobs in a reunited Germany, the German police faced challenges about how to teach the principles of democratic rule of law to officers trained in an autocratic police state.

Border Guards

After 1973, border guard duty was voluntary. Of those who volunteered, most were never charged or held accountable for their actions, asserting that, under East German law, their conduct had been lawful and they therefore could not be held to criminal responsibility. At the trial of the East German border guard who shot and killed Chris Gueffroy, the last to die at the Wall, the guard told the court "at that time I was following the laws and commands of the German Democratic Republic." He was, however, convicted, the judge declaring, "Not everything that is legal is right."

The NVA, National People's Army, or East Germany Army
The NVA was disbanded in 1990. Most facilities were closed, the equipment sold to other countries. Most of the NVA's 36,000 enlisted and noncommissioned officers were let go, including all officers above the rank of lieutenant colonel. Only 3,200 were retained by the Bundeswehr, the German armed forces, after a demotion of at least one rank.

Retired NVA soldiers and officers received minimal pensions after reunification, which left many former NVA officers bitter about their post-reunification treatment. Few were able to find jobs and they were prohibited from appending their NVA rank to their name as a professional title.

The Group of Soviet Forces in Germany, GSFG, Soviet Army in East Germany
August 1994 marked the end of Soviet military presence on German soil, whereby the last of the Soviet soldiers returned to Russia. In 1994, Karlshorst's Sixth Independent Motor Rifle Brigade deactivated and withdrew to Kursk.

The Athletes
Some athletes continued participating and competing in reunited Germany's sports establishment. Some became coaches and trainers. Some athletes filed lawsuits against the German Olympic Committee, and a compensation fund was established for victims of doping. Between 1950 and 1989, some 615 athletes defected to the West.

Escapees and Those Who Attempted
There are no certain statistics about how many people attempted

escape, how many were successful, how many were killed or sent to prison.

Because the East German government covered up so many deaths, there are no definitive statistics on how many were killed, but it is believed that approximately 140 people were killed trying to cross the border in Berlin between 1961 and 1989, and some 1,000 others while trying to cross the border elsewhere or by drowning in the Baltic Sea or in Berlin's Spree River. Some 5,000 others managed to flee, using creative means to make the voyage successfully.

Those who helped escapees were also subject to punishment, facing prison terms or expulsion. Some 50,000 East Germans suffered this fate between 1952 and 1989.

In some cases, would-be escapees were executed in the Soviet Union.

Only a fraction of deaths due to attempted escape have ever been prosecuted or investigated. Today responsibility for the great majority of those deaths remains unaccounted for.

Political Prisoners
All were released.

Some former political prisoners of the Hoheneck Castle are outspoken advocates for former political prisoner victimization, rights, and compensation. Some have written books about their ordeals. The women of Hoheneck face their futures with the support of their unique prison sisterhood, and the women still regularly get together to support one another in Stollberg, in the valley below the prison.

The Dissidents

Ulrike Poppe, Bärbel Bohley, and others like them are credited with helping to lay the intellectual foundation for what was called the Peaceful Revolution of 1989, the fall of the Berlin Wall.

Major Nicholson

At the age of thirty-seven, Major Arthur Nicholson became the last casualty of the Cold War. He was promoted posthumously to lieutenant colonel and is interred at Arlington National Cemetery in Section 7A, Lot 171, not far from the Tomb of the Unknowns.

President Ronald Reagan

Ronald Reagan will be remembered for his hard-line stance against communism and his uncompromising conviction to defeat it. His initiatives on nuclear disarmament and his insistence that the Soviet Union could be defeated rather than simply negotiated with contributed significantly to communism's collapse in Europe. Scholars agree, however, that it was his antinuclear campaign and his leadership and skills as a negotiator that proved pivotal in reaching a negotiated end to the Cold War.

General Secretary of the Soviet Union Mikhail Gorbachev

In 1990 Mikhail Gorbachev won the Nobel Peace Prize for his efforts to end the Cold War. *Time* magazine named him Man of the Year *and* Man of the Decade. Gorbachev continued to serve as general secretary until he resigned as the last head of state of the Soviet Union.

In 1992, he founded the Gorbachev Foundation. Headquartered in San Francisco, the organization aims to contribute to the strengthening and spread of democracy throughout the world. That same year, President Reagan awarded Gorbachev the first-ever Ronald Reagan Freedom Award.

In Germany, Gorbachev remains a hero, where most credit him with laying the foundation that eventually enabled East and West Germany to reunite.

AUTHOR'S NOTE

*T*hroughout East Germany there were tens of thousands, if not millions of people like my relatives, trying to raise a family, work, preserve their dignity, and live life as best they could under the circumstances they were handed. This is the story of just one family, but in some ways it is the story of millions.

The human side of this story is a critical part of East Germany's historical narrative. Depravations at the hands of the system caused many people to rise above and call forth the best of the human spirit. Controlled by a regime with a secret police that manipulated their lives, and devoid of free will, it was often their own spirit and strength of family that helped sustain them. Despite authoritarian rule, lives under repression were ameliorated to a great extent by the love and trust of others in the same situation and, in the end, it was humanity that guided many.

The chronicle of East Germany and the story I tell here is not distant history. It is living memory for millions who lived and experienced it, who are alive today and have their own accounts. Many former East Germans will have had different experiences than those of my family members. They have their stories. This is ours.

And finally, history is rarely black and white. Perspectives vary and the story is often more complicated and nuanced than one can address in a single book. There are some who long for the old days in East Germany, in which, though they were not free, some saw life as simpler, less complicated, devoid of commercialism, and in which low crime rates and cradle-to-grave benefits were guaranteed.

A Note on Research

Research for this book utilized a variety of resources, including archives, fact-finding trips, and interviews. The dates and sequence of events are based on my research and on the best recollections of interviewees over a sixty-five-year period.

Archives and Print Documents

Thanks to a variety of organizations, there is a wealth of publicly available archive material today that helps document the history of the Cold War and East Germany.

These archives include Open Society Archives, RFE/RFL, National Security Archive, Parallel History Project on Cooperative Security, Chronik der Mauer, Imperial War Museum Archives, Stanford University's Hoover Institution, Harvard University's Davis Center for Russian and Eurasian Studies, the Wilson Archives at George Washington University, and the Stasi archives.

I gathered information from family memoirs, letters, and diaries, and reviewed a variety of propaganda materials, including that which the regime forced upon the East German population, including my relatives, to cement ideological perspectives and alter their thinking of world and national events.

Fact-Finding Trips

While living and working in Europe and Eastern Europe, I made research trips to the locations mentioned in the book, often several times, in order to get the lay of the land, and learn more about the region and events that occurred during the Cold War, speaking with residents, historians, and former government officials.

I walked through the village of Schwaneberg, spent time in the tiny but well-cared-for village museum, and spoke to residents who lived there during the Cold War. I saw the schoolhouse where my mother's family lived and where Opa served as teacher and headmaster, and the church where the family went to services before the Soviets occupied the East.

I visited remote Klein Apenburg and saw the house where my grandparents had lived, the yard where Oma had had her garden, and the little church, and even sat on "Opa's resting place" bench.

In Seebenau, I looked over Kallehn's farmstead and walked through what was once his farm fields. Later I went to see the area that had been the East–West border, where a Soviet border guard shot at my mother as she tried to escape, and where a watchtower still stands.

In Stollberg, I toured the Hoheneck Castle and spoke to former political prisoners who were incarcerated there. I also toured the Heidelberg Castle, which is blessed with a much more fortunate legacy.

In Trabitz, I spoke with elderly villagers and saw the schoolhouse where my mother was born and where the family lived when Opa was just beginning his career as a schoolteacher.

Chemnitz today is a vibrant, thriving modern city with few remnants of its communist past, when it was known as Karl Marx City, except for the forty-ton bust of Karl Marx and the former Stasi headquarters building.

Paradise Bungalow is vibrant and alive, and has grown. Today it resembles a small Italian vineyard, with grapevine trellises, a blanket of flowers, and an abundance of vegetables and fruits. The modest patch of land that was once given to Heidi and Reinhard to help grow food for themselves and others is still every bit the powerful symbol of freedom and ingenuity that it was during the Cold War.

I have been to Berlin several times since the fall, once with Cordula. Everything has changed since the Wall fell. We visited the Brandenburg Gate, and Checkpoint Charlie, where little remains from the Cold War days beyond the warning sign telling visitors they are leaving the American sector. The cold, gray passage of entry into the East, which once consisted of border guards and watchtowers, has been replaced with a mockup MP shed, a museum, trinket shops, and vendors who sell communist pins and border guard hats.

I went back to Karlshorst, Unter den Linden, Karl Marx Allee, and other areas our intelligence teams traveled in the East. Leipzigerstrasse today is filled with coffee shops, clothing stores, and life energy, no longer the near-abandoned gray stretch of road that we used to race down, trying to lose our surveillance.

I went back to Potsdam, where the USMLM Mission House was located, to Marzahn, where Cordula trained in the velodrome, and to Clay Headquarters, where I worked in the basement.

Along the way, while living and working in Europe and Eastern Europe, I was also fortunate enough to have been able to engage in many activities that supplemented my research in unexpected ways.

In May 2003, while living in Moscow, I celebrated alongside Russians in Red Square when Paul McCartney gave a historic concert, marking the first time one of the Beatles had played in the former Soviet Union after the band was prohibited from performing there during the Cold War. In 2007 and 2008, on the occasion of the anniversary of German reunification, while living in Prague, I attended receptions at the West German Embassy, where I viewed the lawn on which East Germans fled and took refuge in the last chaotic days before the fall.

Interviews

Living in Central and Eastern Europe gave me excellent access to research for this manuscript. Besides extensive interviews with the family, I spoke with many who lived under communist rule in Eastern Europe during the Cold War, including diplomats from Poland, Czech Republic, Hungary, Romania, Bulgaria, and Russia. In Moscow, I spent time with former Soviets who shared their experiences serving in the Red Army, or who lived as common citizens in the Soviet Union or in East Germany.

Throughout the mid-1980s and after the fall of the Wall, I spoke with dozens of American, British, and French soldiers and civilians who served in Berlin during the Cold War, including USMLM and BRIXMIS staff. I interviewed a dozen former women political prisoners about their experiences at the Hoheneck Castle Women's Prison, and one man who was born and lived in the prison with his mother.

My absolute favorite chats and interviews were with Heidi, Reinhard, and Cordula over fresh-baked bread with homemade jam at Paradise Bungalow.

ACKNOWLEDGMENTS

I would like to thank those who did not hesitate in helping me to tell this story and bring this book to life.

To treasured friends and consummate professionals for their generous reviews and edits, but especially for their friendship:

In the United Kingdom, Major General (Ret.) Peter Williams, former head of the NATO Military Liaison Mission in Moscow, and British Commanders'-in-Chief Mission (BRIXMIS) Tour Officer during the Cold War.

In Canada, Dr. Douglas Parker, Research Professor, Carleton University, and Emeritus Professor of English Literature, Laurentian University, Ontario.

In the United States and in Belarus, Anne Grawemeyer, and in New Zealand and Belarus, Juliet Campbell.

To my friends in the Northern Virginia Writers Group who helped critique the very first draft of this book, especially George Vercessi, Clyde Linsley, Valery Garrett, and Mary Wuest.

A special thank-you to my family for their support throughout the making of this book. To my brother Dr. Albert Willner, who reviewed and gave suggestions to improve both the historical and personal aspects of the story, to my brother Michael for his helpful edits, and to Marcel, Maggy, and Sachi. Special thanks go to my children, Alex, Kim, and Michael, for their support over the years, and especially to Michael for his design and technical guidance.

My mother, Hanna Willner, was my right hand in this process, assisting me every step of the way. Without her support for this project, assistance in translating, deciphering information, and giving me perspective, this work could not have been completed. I am indebted to her for many things, not the least of which is her courage in breaking from the thing she loved most, her family, in order to find and make a better life for her children.

To my father, Eddie Willner, for his wisdom, courage, indomitable spirit, and perseverance to become an American citizen.

I am profoundly grateful to my former East German relatives for helping me to re-create this story. Since the fall of the Berlin Wall and the end of the Cold War, through conversations over bonfires, family parties, walks in the woods, and chats in their homes—it is through their recollections and the places they took me that I came to understand the fuller story. I owe a great debt to my aunts, uncles, cousins—to all my relatives—for their willingness and transparency in sharing their stories, which was not always easy to relive in retelling. Their stories, memories, and personal insights provided a sound foundation on which this story could be built. A special thank-you to those who allowed me to use photos, personal letters and papers, diaries, and memoirs.

I would like to thank Dr. Hope Harrison, Associate Dean for Research and Associate Professor of History and International Affairs at the Elliott School of International Affairs, George Washington University, one of the first Western academics to have access to and translate Cold War documents from Soviet archives in Moscow, who provided a historical review of my manuscript.

I would like to express my most sincere gratitude to the entire William Morrow team at HarperCollins, and most especially to my editors, Emily Krump and Kelly O'Connor, who understood my vision and so professionally ushered this story to the reader. To my German editor, Tanja Ruzicska at Ullstein Propyläen for her assistance and guidance.

I am very fortunate to have an excellent agent in Mackenzie Brady Watson, who took a chance on an unknown author with an unknown story, and then guided and advocated for me brilliantly every step of the way.

Thanks also to the Sülzetal Regional Government and to the village of Schwaneberg, especially Rudolf Wenzel, chairman of the Association for Schwaneberg History, who allowed me to sift through the cherished artifacts and treasures of the village museum and sink into Schwaneberg's Cold War past.

My appreciation to those who aided in my archival and photographic research:

Dagmar Hovestädt and the research team including Marieke Notarp at the BStU Stasi Records Collection and Archives.

To the researchers at the Police History Collection (Polizeihistorische Sammlung), Chronik der Mauer, Berlin Wall Foundation (Stiftung Berliner Mauer), Oxford's Bodleian Library, Harvard's Fung Library, Library of Congress, Sachsen Memorial Foundation (Stiftung Sächsische Gedenkstätten), and Federal Archives in Berlin (Bundesarchiv).

My appreciation to U.S. Army and USMLM personnel, for their fact checking.

I am grateful to others who aided in my research and discovery. While living and working in Germany and in diplomatic posts in Russia, Belarus, and the Czech Republic, I interviewed dozens of diplomats and acquaintances who lived through the Cold War in Eastern European countries.

To Günter Wetzel, who flew his balloon over the border to freedom, and to the many others like him who had the courage to risk it all to have a chance at freedom, and generously shared their stories and provided personal photos. Appreciation also goes to the Czech, German, British, Swiss, and Dutch photographers who documented the Cold War and made their photographs and information available to me, without hesitation and free of charge, in particular Ed and Louise Sijmons, Bettina Rüegger, Ondřej Klauda, Rüdiger Stehn, Reinhard Wolf, Roger Wollstadt, and Mathias Donderer.

A special note to recognize dissidents and political prisoners everywhere who dare to stand up to tyranny and authoritarian leaderships that rule by fear, brutality, and oppression. Many of them often pay the ultimate price in the pursuit of freedom and truth in the name of their fellow countrymen. In that vein, we remember the victims of the East German uprising of 1953, Budapest in 1956, and Prague in 1968. Perhaps they can rest knowing that, in the end, freedom triumphed over authoritarianism.

Special recognition must also be given to the many victims of the Stasi, including the brave women prisoners of the Hoheneck Castle.

I will take this opportunity to thank Mikhail Gorbachev, Ronald Reagan, and world leaders who helped to put an end to Cold War authoritarianism in Europe.

To Cold War warriors, to soldiers and airmen and women in the U.S. and all Allied militaries who played a critical role in helping to shape the course of a new history and whose dedication to service helped ultimately bring an end to the Cold War.

To Opa, my grandfather, who tried his best to stand up for truth and justice. To Oma, who lit the way for me to write this book.

And finally, I wish to thank to my husband, for his research, edits, and reviews, but even more important, for his understanding of my vision. Most of all, I am grateful for his insistence that, above all, I tell the human story within the framework of history.

GLOSSARY

Allied Checkpoints. Checkpoint Alpha: Helmstedt crossing point between West and East Germany; **Checkpoint Bravo:** Dreilinden crossing point between East Germany and West Berlin; **Checkpoint Charlie:** Berlin crossing point between West and East Berlin; crossing point for Allied military personnel and foreigners, including foreign diplomats.

Allies. While this term was originally used in 1945 to identify the Four Powers administering Berlin (United States, Britain, France, Soviet Union), I use it here to denote the Western powers only.

Berlin. Surrounded by East Germany, Berlin was divided into **West Berlin,** itself divided into three Allied sectors: French sector (north), British sector (center), and U.S. sector (south); and **East Berlin,** the Soviet sector.

Communism, Socialism. For the purposes of this book, I use the terms *socialism* and *communism* interchangeably.

Communist Party. The ruling party of East Germany was the Socialist Unity Party, the SED, which was formed in 1946 by forcing the Communist and Social Democratic parties to fuse into one left party, the SED, which essentially became what was known as the Communist Party of East Germany.

East Germans. Though the citizens of the East are most accurately described before the establishment of East Germany in 1949 as Soviet or East Zone Germans, for the purposes of this book and for the sake of ease for the reader, I refer to all German citizens of the East after 1945 as East Germans.

East Germany. Occupied by the Soviets from 1945 to 1989. From 1945 to 1949 known as the Soviet Zone or East Zone. After 1949 known as East Germany.

FDJ. Free German Youth (Freie Deutsche Jugend), the official communist youth organization for ages fourteen to twenty-five.

FRG. Federal Republic of Germany, known in German as BRD: Bundesrepublik Deutschland. West Germany = FRG = BRD.

GDR. German Democratic Republic, known in German as DDR: Deutsche Demokratische Republik. East Germany = GDR = DDR.

JP. Young Pioneers (Junge Pioniere) and Thälmann Pioneers ages six to fourteen.

Jugendweihe. Ceremony marking entry into adulthood and, during Cold War, included swearing loyalty to the regime.

NVA. National People's Army (Nationale Volksarmee), the East German army.

Stasi. Secret police, Ministry for State Security, MfS (Ministerium für Staatssicherheit). Local or regional authorities in the East = Party officials, Stasi, VoPo, or other government or security authority. For the purposes of this book, I use *Stasi, secret police, police, authorities,* and *local authorities* interchangeably because there was no way to delineate what arm of the law was active in any given situation.

U.S. Soviet Sector Flag Tours. Operations in East Berlin.

U.S. Military Liaison Mission (USMLM). Operations in East Germany.

VoPos. The East German Police, the People's Police (Volkspolizei).

West Germans. To keep things simple for the reader, I refer to all West Zone Germans as West Germans, even though that term was not in use until after the establishment of West Germany in 1949.

West Germany. Administered by the Americans, British, and French from 1945 to 1989. From 1945 to 1949 known as the Allied Zone, West Zone, or Bizonia.

BIBLIOGRAPHY

Applebaum, Anne. *Iron Curtain*. New York: Doubleday, 2012.

Auberjonois, Fernand. "East Germany's New Boss Seen as Ulbricht Shadow." *Toledo Blade*, July 18, 1971. https://news.google.com/newspapers?nid=1350&dat=19710718&id=evFOAAAAIBAJ&sjid=1AEEAAAAIBAJ&pg=6615,1281095&hl=en.

Baumgarten, Klaus-Dieter. "Wenn notwendig, dann treffen mit dem ersten Schuß, Rede-Konspekt von DDR-Grenztruppen-Chef Klaus-Dieter Baumgarten, 9. Juli 1982." Chronik der Mauer, July 9, 1982. http://www.chronik-der-mauer.de/material/178862/wenn-notwendig-dann-treffen-mit-dem-ersten-schuss-rede-konspekt-von-ddr-grenztruppen-chef-klaus-dieter-baumgarten-9-juli-1982.

Beckhusen, Robert. "New Documents Reveal How a 1980's Nuclear War Scare Became a Full-Blown Crisis." *Wired*, May 16, 2013. http://www.wired.com/2013/05/able-archer-scare/.

Beevor, Antony. *Berlin*. London: Viking, 2002.

"The Berlin Duty Train." U.S. Army Transportation Museum, May 15, 2015. http://www.transchool.lee.army.mil/museum/transportation%20museum/bertrain.htm.

Berndt, Peter. "Views of Sport; East Germany: From Both Sides Now." *New York Times*, December 10, 1989. http://www.nytimes.com/1989/12/10/sports/views-of-sport-east-germany-from-both-sides-now.html.

Birch, Adrian. "Iron Curtain's 100,000 Dead." *Independent*, October 27, 2001. http://www.paulbogdanor.com/left/eastgermany/dead.html.

Birch, Douglas. "The U.S.S.R. and U.S. Came Closer to Nuclear War Than We Thought." *The Atlantic*, May 28, 2013. http://www.theatlantic.com/international/archive/2013/05/the-ussr-and-us-came-closer-to-nuclear-war-than-we-thought/276290/.

"Book of Reports Presented at an Official Conference Published by Dietz Verlag—'Oral Agitation.'" German Propaganda Archive, Calvin College, 1984. http://research.calvin.edu/german-propaganda-archive/mund.htm.

Burant, Stephen. *East Germany*. Washington, DC: Federal Research Division, Library of Congress, Country Study, 1987. http://www.country-data.com/cgi-bin/query/r-5015.html.

Burkhardt, Heiko. "Berlin, Berlin Wall and Germany Photographs and Pictures." Dailysoft.com, 2015. http://www.dailysoft.com/berlinwall/photographs/index.htm.

Byrne, Malcolm. "Uprising in East Germany 1953—Shedding Light on a Major Cold War Flashpoint." National Security Archive Electronic Briefing Book No. 50, June 15, 2001. George Washington University National Security Archive. http://nsarchive.gwu.edu/NSAEBB/NSAEBB50/.

Campbell, Bradley. "During the Cold War, Buying People from East Germany Was Common Practice." PRI the World, June 11, 2014. http://www.pri.org/stories/2014-11-06/during-cold-war-buying-people-east-germany-was-.

"Chronicle of the Berlin Wall 1961." Chronik der Mauer. http://www.chronik-der-mauer.de/en/chronicle/#anchoryear1961.

Church, George. "Freedom! The Berlin Wall." Time, November 20, 1989. http://time.com/3558854/freedom-the-berlin-wall/.

"CIA Current Intelligence Review Analyzing the Communist 'New Look in East Germany' and 'Recent Unrest in Eastern Europe (Declassified).'" June 17, 1953. Wilson Center Digital Archive. http://digitalarchive.wilsoncenter.org/document/111320.

"The City of Spies (and the Death of Major Nicholson)." NBC News, March 1985. https://www.youtube.com/watch?v=ISp63YmvvMw.

Clark, Zsuzsanna. "Oppressive and Grey? No, Growing Up Under Communism Was the Happiest Time of My Life." Daily Mail, October 17, 2009. http://www.dailymail.co.uk/news/article-1221064/Oppressive-grey-No-growing-communism-happiest-time-life.html.

Clay, Lucius D. "Berlin." Foreign Affairs 41, no. 1 (1962): 47.

"Cold War History." Wilson Center Digital Archive. http://digitalarchive.wilsoncenter.org/theme/cold-war-history.

"Cold War Origins." Wilson Center Digital Archive. http://digitalarchive.wilsoncenter.org/collection/27/cold-war-origins.

"Communism: The Rise of the Other Germany." Time, October 1, 1971.

"Communists: The Vopos." Time, June 23, 1952.

Connolly, Kate. "Stasi File Details Plans for Riot During 1988 Michael Jackson Concert." Guardian, July 30, 2009. http://www.theguardian.com/music/2009/jul/30/michael-jackson-berlin-wall-germany.

Cooley, John K. Unholy Wars. London: Pluto Press, 2000.

Crane, Keith. "East Germany's Military: Forces and Expenditures." Rand Arroya Center, October 1989. https://www.rand.org/content/dam/rand/pubs/reports/2007/R3726.1.pdf.

Crossland, David. "Painful Memories of an East German Gulag: 'I Thought I Was in a Nazi Movie.'" Spiegel Online, May 6, 2009. http://www.spiegel.de/international/germany/painful-memories-of-an-east-german-gulag-i-thought-i-was-in-a-nazi-movie-a-623008.html.

Currey, Andrew. "Piecing Together the Dark Legacy of East Germany's Secret Police." Wired, January 18, 2008. http://archive.wired.com/politics/security/magazine/16-02/ff_stasi?currentPage=all.

"Death Strip: Berlin Pays Tribute to Last Person Shot Crossing Wall." *Spiegel Online*, February 6, 2009. http://www.spiegel.de/international/germany/death-strip-berlin-pays-tribute-to-last-person-shot-crossing-wall-a-605967.html.

Dempsey, Judy. "East German Shoot-to-Kill Order Is Found." *New York Times*, August 13, 2007. http://www.nytimes.com/2007/08/13/world/europe/13germany.html?pagewanted=print&_r=0.

"Die vermeintliche oder tatsächliche Pleite der DDR." Ddr-wissen.de, undated. http://www.ddr-wissen.de/wiki/ddr.pl?Pleite.

Donovan, Barbara. "Radio Free Europe Background Report—Honecker Speaks Out on Soviet Reforms." Osaarchivum.org, October 15, 1987. http://www.osaarchivum.org/greenfield/repository/osa:f0e78a8a-cbae-4870-ab6f-72ac17101589.

———. "Radio Free Europe Report—Honecker Continues to Reject Reform." Osaarchivum.org, May 8, 1987. http://osaarchivum.org/files/holdings/300/8/3/text/27-4-97.shtml.

Drakulic, Slavenka. *How We Survived Communism and Even Laughed*. New York: Harper Perennial, 1993.

Dulles, Eleanor Lansing. *Berlin: The Wall Is Not Forever*. Chapel Hill: University of North Carolina Press, 1967.

"East German Communism vs. the Perils of the Free West." Deano's Travels, March 29, 2013. https://deanoworldtravels.wordpress.com/2013/03/29/east-german-communism-vs-the-perils-of-the-free-west/.

"East German Ministry of State Security, 'New Methods of Operation of Western Secret Services.'" November 1958. Wilson Center Digital Archive. http://digitalarchive.wilsoncenter.org/document/118653.pdf?v=4ad0b97faadcd76d0a5d940bf7d4eab8.

"East German Uprising." Wilson Center Digital Archive. http://digitalarchive.wilsoncenter.org/collection/35/east-german-uprising.

"East Germany: Alarm." *Time*, February 18, 1957.

"East Germany: Desolate & Desperate." *Time*, August 4, 1961.

"East Germany: Exile for Heretics." *Time*, October 3, 1977.

"East Germany: Intolerable Conditions." *Time*, February 23, 1962.

"East Germany: Making the Best of a Bad Situation." *Time*, October 17, 1969.

"East Germany: The Vanishing Intellectuals." *Time*, September 1, 1958.

"East Germany: They Have Given Up Hope." *Time*, December 6, 1963.

"East-West: Life along the Death Strip." *Time*, September 15, 1980.

Einhorn, Barbara. *Cinderella Goes to Market*. London: Verso, 1993.

Emerson, Steven. "Where Have All His Spies Gone?" *New York Times*, August 12, 1990. http://www.nytimes.com/1990/08/12/magazine/where-have-all-his-spies-gone.html.

"Escapes from East Germany." November 4, 1961. Vera and Donald Blinken Open Society Archives at Central European University. http://catalog.osaarchivum.org/catalog/osa:57a3da27-abf0-4003-bb35-e0c07baf788e.

"Escaping the East by Any Means—A Look at Ten of the Most Dramatic Escapes." Al Jazeera, November 13, 2009. http://www.aljazeera.com/focus/2009/10/200910793 416112389.html.

"Euromissiles Crisis." Wilson Center Digital Archive. http://digitalarchive.wilsoncenter .org/collection/38/euromissiles-crisis.

Feffer, John. "Remembering the Calm Life Under Communism." *Huffington Post*, February 12, 2013. http://www.huffingtonpost.com/john-feffer/remembering-the-calm-life_b_2671955.html.

Fischer, Benjamin. "A Cold War Conundrum: The 1983 Soviet War Scare." Central Intelligence Agency Library, July 7, 2008,

Fisher, Marc. *After the Wall*. New York: Simon & Schuster, 1995.

"Freie Deutsche Jugend Mitglied Im Weltbund Der Demokratischen Jugend (WBDJ) Seit 1948." Freie Deutsche Jugend. http://www.fdj.de/GESCHI.html.

Fulbrook, Mary. *Anatomy of a Dictatorship*. New York: Oxford University Press, 1995.

Funder, Anna. *Stasiland*. London: Granta. 2003.

Furlong, Ray. "Honecker's Cars Under the Hammer." BBC News, November 4, 2003. http://news.bbc.co.uk/2/hi/europe/3240773.stm.

"Further Loosening of East German Straight-jacket: Economic Journal Warned for Too Rigid Approach." October 30, 1958. Radio Free Europe/Radio Liberty Research Institute, Europeana Collections. http://www.europeana.eu/portal/ record/2022062/10891_osa_72fb15e3_fa5d_4ab9_a1dd_8a80fdc7a810.html.

Gaddis, John Lewis. *The Cold War*. New York: Penguin Press, 2005.

"GDR Official Brochure Published in Defense of the Berlin Wall (What You Should Know About the Wall)." 1962. German Propaganda Archive, Calvin College. http:// research.calvin.edu/german-propaganda-archive/wall.htm.

"GDR Official Pamphlet Published by Bezirk Karl Marz Stadt Department for Propaganda Agitation)—'Farmer Arnold and His Relationship to Socialism.'" 1960. German Propaganda Archive, Calvin College. http://research.calvin.edu/german-propaganda-archive/arnold.htm.

"GDR Official Pamphlet (*Sozialistische Bildungshefte*) published by Dietz Verlag—'The Tasks of Party Propaganda.'" 1950. German Propaganda Archive, Calvin College. http://research.calvin.edu/german-propaganda-archive/sedprop.htm.

Grant, R. G. *The Rise and Fall of the Berlin Wall*. New York: Mallard Press, 1991.

Gray, William Glenn. *Germany's Cold War*. Chapel Hill: University of North Carolina Press, 2003.

Gross, Daniel. "The Man Who Helped Dig a Secret Tunnel Under the Berlin Wall." *Slate*, March 17, 2003. http://www.slate.com/blogs/the_eye/2014/03/07/the_ history_of_tunnel_57_a_secret_escapeway_under_the_berlin_wall.html.

Haines, Gavin. "East Germany's Trade in Human Beings." *BBC Magazine*, November 6, 2014. http://www.bbc.com/news/magazine-29889706.

Harrison, Hope. "After the Berlin War: Memory and the Making of the New Germany, 1989 to the Present." Unpublished manuscript, 2016.

———. *Driving the Soviets up the Wall*. Princeton, NJ: Princeton University Press, 2005.

———. *Ulbrichts Mauer*. Berlin: Ullstein Propyläen Verlag, 2011.

Hertle, Hans-Hermann, and Maria Nooke. *The Victims at the Berlin Wall, 1961–1989: A Biographical Handbook*. Berlin: Ch. Links Verlag, 2011.

Hignett, Kelly. "'Endut! Hoch Hech!': Confronting Stereotypes About Everyday Life in Communist Eastern Europe." The View East, July 27, 2015. https://thevieweast .wordpress.com/2015/07/27/endut-hoch-hech-confronting-stereotypes-about-everyday-life-in-communist-eastern-europe/.

———. "Paula Kirby on Life in the GDR." The View East, February 14, 2014. https:// thevieweast.wordpress.com/2014/02/14/paula-kirby-on-life-in-the-gdr/.

"History of the United States Military Liaison Mission (USMLM)." USMLM Association. http://usmlm.us/history/.

Hoffman, Adolph. "Part 6: Berlin and the Two Germanies (East Germany During the Wall)." http://art.members.sonic.net/unify90/ber6.html.

Holbrook, James R. *Potsdam Mission*. Carmel, IN: Cork Hill Press, 2005.

Honecker, Erich. "The FDJ—A Reliable Partner in the Struggle of Our Party." Speech to the FDJ published in *Nueus Deutschland*, May 13–14, 1989. http://research.calvin .edu/german-propaganda-archive/fdj.htm.

———. "Party and Revolutionary Young Guard Firmly Allied." Speech to the FDJ published in *Nueus Deutschland*, May 13–14, 1989. http://research.calvin.edu/ german-propaganda-archive/fdj.htm.

"How the Stasi Defamed Günter Litfin, the First Victim of the Wall." Federal Commissioner for the Records of the State Security Service of the Former German Democratic Republic. http://www.bstu.bund.de/EN/PressOffice/Topics/litfin .html?nn=3625808.

Huggler, Justin. "An Audacious Dig for Freedom Beneath the Berlin Wall." *Telegraph*, October 5, 2014. http://www.telegraph.co.uk/news/worldnews/europe/germany/ 11140109/An-audacious-dig-for-freedom-beneath-the-Berlin-Wall.html.

"Implications of Recent Soviet Military Political Activities." Declassified Special National Intelligence Estimate, May 18, 1984. George Washington University National Security Archive. http://nsarchive.gwu.edu/NSAEBB/NSAEBB428/docs/6.Implications%20 of%20Recent%20Soviet%20Military-Political%20Activities.pdf.

"Intelligence Operations in the Cold War." Wilson Center Digital Archive. http://digi talarchive.wilsoncenter.org/collection/45/intelligence-operations-in-the-cold-war/4.

"Iron Curtain Kid—A Boy's Life in the GDR." http://www.ironcurtainkid.com/.

Johnson, Marguerite. "West Germany Spies, Spies and More Spies, Kohl Moves to Limit the Damage as the Scandal Widens." *Time*, September 9, 1985. http://content .time.com/time/subscriber/article/0,33009,959796,00.html.

Johnson, Molly Wilkinson. *Training Socialist Citizens*. Leiden: Brill, 2008.

Jones, John Penman. "Voices Under Berlin: Night Train to Berlin." Drawn from "Night Train to Berlin," *Army in Europe*, August 1969, pp. 26–27. http://www.voicesunder berlin.com/BerlinTravel/NightTraintoBerlin.html.

Jones, Nate. "1983 War Scare: 'The Last Paroxysm' of the Cold War." Parts 1–3. George Washington University National Security Archive, May 16, 2013. http://nsarchive .gwu.edu/NSAEBB/NSAEBB426/; http://nsarchive.gwu.edu/NSAEBB/ NSAEBB427/; http://nsarchive.gwu.edu/NSAEBB/NSAEBB428/.

Jones, Tamara. "Update: Honecker a Prisoner of the Past: Three Countries Are Deadlocked over the Fate of the Ailing Former East German Leader." *Los Angeles Times*, February 2, 1992. http://articles.latimes.com/1992-02-22/news/mn-1950_1_east-german.

Kirchner, Kurt. "Arbeitet So, Wie Die Agitatoren Wilk Und Porsch!" *Notizbuch Des Agitators* 3 (1953): 31–35.

Klee, Ernst. "Psychiatrie in der DDR—eine erste Bestandsaufnahme: Wecken um halb vier." *Zeit Online*, June 28, 1991. http://www.zeit.de/1991/27/wecken-um-halb-vier.

Kloth, Hans Michael. "East German by Design: The ABCs of Communist Consumer Culture." *Spiegel Online International*, June 23, 2009. http://www.spiegel.de/ international/germany/east-german-by-design-the-abcs-of-communist-consumer-culture-a-631992.html.

Koehler, John O. *Stasi*. Boulder, CO: Westview Press, 1999.

Kovaly, Heda Margolius. *Under a Cruel Star*. Trans. Helen Epstein. Cambridge, MA: Plunkett Lake Press, 1986.

Kramer, Jörg. "Escape via Elevator Shaft: East Germany's 'Traitor Athletes' Tell Their Stories." *Spiegel Online*, July 22, 2011. http://www.spiegel.de/international/germany/ escape-via-elevator-shaft-east-germany-s-traitor-athletes-tell-their-stories-a-775370 .html.

Kusin, Vladimir V. "Gorbachev in East Germany." "Shortage of Raw Materials in East Germany." Radio Free Europe/Radio Liberty Research Institute Report. Vera and Donald Blinken Open Society Archives at Central European University. October 19, 1989. http://catalog.osaarchivum.org/catalog/osa:bda3a8f2-5282-4b43-9ccd-314c57aa213e.

Lander, Mark. "Stollberg Journal; to Onetime Political Inmates, This Jail Was No Vacation." *New York Times*, August 19, 2004. http://www.nytimes.com/2004/08/19/world/ stollberg-journal-to-onetime-political-inmates-this-jail-was-no-vacation.html.

Larson, Erik. *In the Garden of Beasts*. New York: Crown, 2011.

Latotzky, Alex. "A Childhood Behind Barbwire." http://alex.latotzky.de/Estart.htm.

Lavigne, Marie. *International Political Economy and Socialism*. Cambridge: Cambridge University Press, 1991.

Lewis, Ben. *Hammer and Tickle*. London: Phoenix, 2009.

Lukas, Karin. "East Germany's October 'Spring': A German Journalist Recalls Leipzig in 1989 and the Protests That Led to the Fall of East Germany." Al Jazeera, October 9, 2014. http://www.aljazeera.com/indepth/opinion/2014/10/east-germany-october-spring-201410813025891431.html.

Lynch, Denni. "The Berlin Wall, 1961–1989: A Timeline of a Divided Germany." *International Business Times*, October 11, 2014. http://www.ibtimes.com/berlin-wall-1961-1989-timeline-divided-germany-1721012.

Maaz, Hans-Joachim. *Behind the Wall*. New York: Norton, 1995.

Major, Patrick. "Holding the Line: Policing the Open Border." In *Behind the Berlin Wall: East Germany and the Frontiers of Power*. New York: Oxford University Press, 2010.

Mars, Roman. *Tunnel 57*. Episode 104 (video). 99PercentInvisible.org. http://99percentinvisible.org/episode/tunnel-57/.

McKenna, David. *East Germany*. New York: Chelsea House, 1988.

"The Meaning of Being a Soldier." German Propaganda Archive, Calvin College. http://research.calvin.edu/german-propaganda-archive/soldat.htm.

Meyer, Michael. *The Year That Changed the World*. New York: Scribner, 2009.

"Military Parade Celebrating the 40th Anniversary of the GDR (October 7, 1989)." German History Through Documents and Images. http://germanhistorydocs.ghi-dc.org/sub_image.cfm?image_id=3025.

"Minutes No. 49 of the Meeting of the SED Politburo." Wilson Center Digital Archive, November 7, 1989. http://digitalarchive.wilsoncenter.org/document/113041.

Molloy, Peter. *The Lost World of Communism*. London: BBC, 2009.

"Nachdrückliche Warnung note der Britischen Regierung an Moskau—Die Luftkorridore." Chronik der Mauer. http://www.chronik-der-mauer.de/material/180307/antwortnote-der-westmaechte-auf-die-note-der-sowjetunion-vom-23-august-1961.

Naimark, Norman. "About 'The Russians' and About Us: The Question of Rape and Soviet-German Relations in the Soviet Zone." Washington, DC: National Council for Eurasian and East European Research, 1991. http://www.ucis.pitt.edu/nceeer/pre1998/1991-802-14-2-Naimark.pdf.

———. *The Russians in Germany*. Cambridge, MA: Belknap Press of Harvard University Press, 1995.

Nettl, Peter. "German Reparations in the Soviet Empire." *Foreign Affairs* 29, no. 2 (1951): 300.

"Nie Wieder ein Krieg von Deuscher Erde!" *Volksstimme* (Magdeburg, East Germany), June 23, 1966.

"1956 Polish and Hungarian Crises." Wilson Center Digital Archive. http://digitalarchive.wilsoncenter.org/collection/9/1956-polish-and-hungarian-crises.

"Памятный знак жертве 'холодной войны.'" Group of Soviet Forces Germany, October 2, 2013. http://www.gsvg.ru/2ww/93-pamyatnyy-znak-zhertve-holodnoy-voyny.html.

"Official GDR Pamphlet 'Notiz Buch Des Agitators' Published by Socialist Unity Party's Agitation Department, Berlin District—'He Who Leaves the GDR Joins the Warmongers.'" 1955. German Propaganda Archive, Calvin College. http://research .calvin.edu/german-propaganda-archive/notiz3.htm.

Oltermann, Phillip. "Surfboards and Submarines: The Secret Escape of East Germans to Copenhagen." *Guardian*, October 17, 2014. http://www.theguardian.com/cities/2014/ oct/17/surfboards-and-submarines-the-secret-escape-of-east-germans-to-copenhagen.

"Our Five Year Plan for Peaceful Reconstruction." 1952. German Propaganda Archive, Calvin College. http://research.calvin.edu/german-propaganda-archive/5yrplan.htm.

Painton, Frederick. "Eastern Europe Communism's Old Men—Gorbachev Tries to Introduce Change to the Aging Party Bosses." *Time*, April 28, 1986. http://content .time.com/time/subscriber/article/0,33009,961281,00.html.

Pankau, Matthias. "How Bodo Strehlow Paid for His Attempt to Escape from East Germany with Solitary Confinement." *German Times for Europe*, November– December 2010. http://www.german-times.com/index.php?option=com_content&t ask=view&id=38373&Itemid=182.

"The Pershing Missile System and the Cold War." Cold War Museum. http://www .coldwar.org/articles/50s/pershing_missiles.asp.

Pieck, Wilhelm. "Zehn Yahre Deutsche Demokratische Republic." German Propaganda Archive, Calvin University. http://research.calvin.edu/german-propaganda-archive/ gdrmain.htm.

"Plea from 'the Housewives in the Soviet Zone.'" Gustrow, Germany, 1946. Warwick Cold War Archives. http://www2.warwick.ac.uk/services/library/mrc/explore further/images/coldwar/.

Portes, Richard. "East Europe's Debt to the West: Interdependence Is a Two-Way Street." *Foreign Affairs* 55, no. 4 (July 1977): 751.

"Post-War East Germany." Excerpt from Glenn E. Curtis. *East Germany: A Country Study*. Washington, DC: Federal Research Division of the Library of Congress, 1992. http://www.shsu.edu/~his_ncp/EGermPW.html.

"The Price of Gold: The Legacy of Doping in the GDR." *Spiegel Online International*, August 17, 2009. http://www.spiegel.de/international/germany/the-price-of-gold- the-legacy-of-doping-in-the-gdr-a-644233.html.

"Protokoll Nr. 45/61 der Sitzung des Politbüros des Zentralkomitees der SED am Dienstag, dem 22. August 1961, im Sitzungssaal des Politbüros." Chronik der Mauer. http://www.chronik-der-mauer.de/index.php/de/Start/Detail/id/593839/ page/27.

Reagan, Ronald. "I'm Convinced That Gorbachev Wants a Free-Market Democracy." *New York Times*, June 12, 1990. http://www.nytimes.com/1990/06/12/opinion/i-m- convinced-that-gorbachev-wants-a-free-market-democracy.html.

Roberts, Andrew. "Stalin's Army of Rapists: The Brutal War Crime That Russia and Germany Tried to Ignore." *Daily Mail Online*, October 24, 1988. http://www.daily

mail.co.uk/news/article-1080493/Stalins-army-rapists-The-brutal-war-crime-Russia-Germany-tried-ignore.html.

Robson, Jeff. "'He Who Has the Youth, Has the Future': Youth and the State in the German Democratic Republic." History Matters, undated. http://historymatters .appstate.edu/sites/historymatters.appstate.edu/files/youthstateGDR_000.pdf.

Rodden, John. "Creating 'Textbook Reds.'" Society 42, no. 1 (2004): 72–78.

———. Repainting the Little Red Schoolhouse. New York: Oxford University Press, 2002.

Rosen, Armin. "A Declassified CIA Paper Shows How Close the U.S. and the Soviets Really Came to War in 1983." Business Insider, September 18, 2014. http://www .businessinsider.com/how-close-the-us-and-the-soviets-came-to-war-in-1983-2014-9.

"The Rules of the Thälmann Pioneers." German Propaganda Archives, Calvin College. http://research.calvin.edu/german-propaganda-archive/tp.htm.

Sauer, Heiner, and Hans-Otto Plumeyer. Der Salzgitter Report. Esslingen: Bechtle, 1991.

Schaefer, Bernd, and Christian Nuenlist. "NATO's 'Able Archer 83' Exercise and the 1983 Soviet War Scare." Parallel History Project on Cooperative Security, November 6, 2003. http://www.php.isn.ethz.ch/collections/colltopic.cfm?lng=en&id= 16431&navinfo=15296.

———. "Stasi Intelligence on NATO, 1969–1989." Parallel History Project on Cooperative Security, November 2003.

Schlosser, Nicholas J. "The Berlin Radio War: Broadcasting in Cold War Berlin and the Shaping of Political Culture in Divided Germany 1945–1961." Ph.D. diss., University of Maryland, 2008.

Schmid, Heinz D. Fragen an Die Geschichte. 4th ed. Frankfurt am Main: Hirschgraben.

Schmidt, Wolfhard. "The Right Word at the Right Time: On the Work of the Agitator." Radar 4 (1988). German Propaganda Archive, Calvin College. http://research.calvin .edu/german-propaganda-archive/radar1.htm.

Schöne, Jens. The GDR: A History of the Workers' and Peasants' State. Trans. Simon Hodgson. Berlin: Berlin Story Verlag, 2015.

Schultke, Dietmar. Keiner kommt Durch Die Geschichte der innerdeutschen Grenze und der Berliner Mauer. Berlin: Aufbau, 1999.

"Senator Joseph McCarthy, McCarthyism, and the Witch Hunt." Cold War Museum. http://www.coldwar.org/articles/50s/senatorjosephmccarthy.asp.

Shapiro, Susan, and Ronald Shapiro, The Curtain Rises. Jefferson, NC: McFarland, 2004.

"Shortage of Raw Materials in East Germany." Radio Free Europe/Radio Liberty Research Institute Report. August 21, 1951. Vera and Donald Blinken Open Society Archives at Central European University. http://www.osaarchivum.org/greenfield/ repository/osa:cc99d470-13f4-4c89-ae29-966ef47856c2.

Silberstein, Gerard E. "Uprising in East Germany: The Events of June 17, 1953." History: Reviews of New Books 1, no. 2 (1972): 41.

Smyser, W. R. *From Yalta to Berlin: The Cold War Struggle over Germany.* New York: St. Martin's Press, 1999.

Sokolovskii, V., and L. Govorov. "Report from V. Sokolovskii and L. Govorov in Berlin to N.A. Bulganin." June 17, 1953. Wilson Center Digital Archive. http://digital archive.wilsoncenter.org/document/112449.pdf?v=ebe3b90346435974c 3f338850a96ed74.

Steiner, André. *The Plans That Failed: An Economic History of the GDR.* Trans. Ewald Osers. New York: Berghahn Books, 2010.

Suess, Walter, and Douglas Selvage. "KGB/Stasi Cooperation." Cold War International History Project and the Office of the Federal Commissioner for the Stasi Records, October 27, 2012. http://www.wilsoncenter.org/publication/e-dossier-no-37-kgbstasi-cooperation?gclid=CJXWkqqs88YCFVTMtAodM2IG7g.

Swain, Geoff, and Nigel Swain. *Eastern Europe Since 1945.* New York: St. Martin's Press, 1993.

"Swim for Your Life out of the GDR!" YouTube video, undated. https://www.youtube .com/watch?v=K7CWajaOx4E.

Taylor, Fredrick. *The Berlin Wall: A World Divided, 1961–1989.* New York: HarperCollins, 2006.

Turner, Henry Ashby. *Germany from Partition to Reunification.* New Haven, CT: Yale University Press, 1992.

———. *The Two Germanies Since 1945.* New Haven, CT: Yale University Press, 1987.

"Two Berlins: A Generation Apart." *National Geographic,* January 1982, 3–51.

"Two Pledges for the Jugendweihe (1955/1958)." Translated by Thomas Dunlap. German History Through Documents and Images. http://germanhistorydocs.ghi-dc .org/sub_document.cfm?document_id=4573.

"Uchtspringe (Landesheilanstalt Uchtspringe)." http://www.uvm.edu/~lkaelber/ children/uchtspringe/uchtspringe.html.

Uhl, Matthias, Christian Nuenlist, and Anna Locher. "The 1961 Berlin Crisis and Soviet Preparations for War in Europe." Parallel History Project, December 4, 2003. http://www.php.isn.ethz.ch/collections/colltopic.cfm?lng=en&id=16161.

Ulbricht, Walter. "Letter from Ulbricht to Khrushchev on Closing the Border Around West Berlin." September 15, 1961. Wilson Center Digital Archive. http://digital archive.wilsoncenter.org/document/116212.

Ungerleider, Steven. *Faust's Gold.* New York: Thomas Dunne Books, 2001.

"Unit History: United States Military Mission to the Commander in Chief, Group of Soviet Forces in Germany." 1985. U.S. Army, declassified document.

United States Military Liaison Mission. "Cold War Timeline." http://usmlm.us/.

"Upheaval in the East; Honecker and Wife Hounded from New House by a Mob." *New York Times,* March 3, 1990. http://www.nytimes.com/1990/03/25/world/upheaval-in-the-east-honecker-and-wife-hounded-from-new-house-by-a-mob.html.

USSR Council of the Ministers. "Measures to Improve the Health of the Political

Situation in the GDR." June 2, 1953. Wilson Center Digital Archive. http://
digitalarchive.wilsoncenter.org/document/110023.

"U.S. Vehicle Is Hit in the East." *New York Times*, July 17, 1985. http://www.nytimes
.com/1985/07/17/world/us-vehicle-is-hit-in-east-germany.html.

"The Wall That Defined Us." *Time*, November 5, 1999.

"Walter Ulbrichts 'Dringender Wunsch.'" Bundeszentrale Für Politische Bildung,
September 1, 2012. http://www.bpb.de/geschichte/deutsche-einheit/deutsche-
teilung-deutsche-einheit/52213/walter-ulbrichts-dringender-wunsch.

"Wer Ist Schuld Am Bau Und Am Fortbestand Der Mauer Zwischen DDR Und
BRD?" Der-demokrat.de. http://www.der-demokrat.de/.

"What You Should Know About the Wall." German Propaganda Archive, Calvin
College, 1962. http://research.calvin.edu/german-propaganda-archive/wall.htm.

Williams, Peter. "Being Detained by the Soviets Could Be Fun—A BRIXMIS
Detention near Wittenberg 27–28 August 1981." Parallel History Project, April
2007. http://www.php.isn.ethz.ch/collections/colltopic.cfm?lng=en&id=27752.

Williams, Peter, and Leo Niedermann, eds. "British Commanders-in-Chief Mission to
the Soviet Forces in Germany (BRIXMIS): Photographs and Documents." Parallel
History Project, April 5 2007. http://www.php.isn.ethz.ch/collections/colltopic
.cfm?lng=en&id=27752.

Winkler, Christopher, Anna Locher, and Christian Nuenlist, eds. "Between Conflict
and Gentleman's Agreement." Parallel History Project, July 1, 2005. http://www
.php.isn.ethz.ch/collections/colltopic.cfm?lng=en&id=14644.

Wölbern, Jan Philipp. *Der Häftlingsfreikauf aus der DDR 1962/62–1989*. Göttingen:
Vandenhoeck & Ruprecht, 2014.

"World: Wall of Shame." *Time*, August 31, 1962.

"Zeitgeschichte: Tödliche Schüsse ohne Vorwarnung." *Spiegel Online*, March 20,
2005. http://www.spiegel.de/panorama/zeitgeschichte-toedliche-schuesse-ohne-
vorwarnung-a-347432.html.

IMAGE CREDITS

Grateful acknowledgment is made to the following for the use of the photographs that appear throughout the text:

Courtesy of the Willner family (pages xx, 10–11, 15, 26, 32, 90, 104–5, 124, 126, 132, 158, 166, 180–82, 184, 187, 200–201, 242, 285, 300, and 339)

Courtesy of Heimatverein Schwaneberg e.V. (page 2)

Courtesy of Bundesarchiv, Bild 183-13735-0006/Photo Walter Heilig (page 27)

Courtesy of Universal History Archive/Getty Images (page 96)

Courtesy of Keystone/Getty Images (page 147)

Courtesy of Bundesarchiv, Bild 183-R1220-401 (page 189)

Courtesy of Bundesarchiv, Bild 183-1984-1126-312 (page 193)

Courtesy of Rüdiger Stehn (page 253)

Courtesy of BStU (pages 261, 264, and 265 top)

Courtesy of BRIXMIS Association (pages 265 bottom, 271, and 275)

Photograph by author (page 272)

Courtesy of Eike C. Radewahn (page 280)

Courtesy of Thomas Hoepker/Magnum Photos (page 291)

INDEX

Note: *Italicized* page numbers indicate photographs.

ABOUT THE AUTHOR

NINA WILLNER is a former U.S. Army intelligence officer who served in Berlin during the Cold War. Following a career in intelligence, Nina worked in Moscow, Minsk, and Prague, promoting human rights, children's causes, and the rule of law for the U.S. government, nonprofit organizations, and a variety of charities. She currently lives in Istanbul, Turkey. *Forty Autumns* is her first book.